giving
nature
a home

Ducks and Geese

Marianne Taylor

GW00685615

BLOOMSBURY WILDLIFE
LONDON · OXFORD · NEW YORK · NEW DELHI · SYDNEY

BLOOMSBURY WILDLIFE
Bloomsbury Publishing Plc
50 Bedford Square, London, WC1B 3DP, UK

BLOOMSBURY, BLOOMSBURY WILDLIFE and the Diana logo are trademarks of
Bloomsbury Publishing Plc

First published in the United Kingdom, 2020

A catalogue record for this book is available from the British Library

Library of Congress Cataloguing-in-Publication data has been applied for

ISBN: PB: 978-1-4729-7164-7; ePDF: 978-1-4729-7165-4; eBook: 978-1-4729-7166-1

2 4 6 8 10 9 7 5 3 1

Design by Susan McIntyre
Printed and bound in India by Replika Press Pvt. Ltd.

MIX
Paper from
responsible sources
FSC® C016779

To find out more about our authors and books visit www.bloomsbury.com and
sign up for our newsletters

giving
nature
a home

Published under licence from RSPB Sales Limited to raise awareness of the RSPB (charity registration
in England and Wales no 207076 and Scotland no SC037654).

For all licensed products sold by Bloomsbury Publishing Limited, Bloomsbury Publishing Limited
will donate a minimum of 2% from all sales to RSPB Sales Ltd, which gives all its distributable profits
through Gift Aid to the RSPB.

Contents

Meet the Ducks and Geese

Whether you are wandering around the pond in your local park or exploring a vast estuary on some wild and remote stretch of coastline, you will encounter some of the UK's many species of ducks and geese. Charming, comical and breathtaking in equal measure, these are the best known and most beloved of our waterbirds, although some of the British species are far from familiar. Many of those that we see in winter have come here from the high Arctic, escaping colder weather in their breeding grounds; some are true seabirds and rarely seen inland.

Ducks and geese are mostly medium or large birds that are adapted to live in and around water for part or most of their lives. They are plump-bodied, with dense, water-repelling plumage, they have short, strong legs, and their front three toes are connected by webbing, to allow them to swim with strokes of their feet underwater. They typically have medium-length, flattened, blunt-tipped bills. Their wings are usually powerful but relatively small for their body size, and while some can fly very fast, their flight is not very energy efficient and a few species cannot fly at all.

Opposite: White-fronted Geese are native birds that visit the UK in winter. They are easily identified by the distinctive black bars on their belly and the large white patch on their head.

Left: Long-tailed Ducks nest in the Arctic and visit our northern coasts in winter.

Right: Drake (male) Mallards acquire their resplendent breeding plumage in early winter.

Below: Tundra Bean Geese migrate from Russia to central Europe in winter, though only small numbers reach the UK.

Most ducks and geese are very sociable – not only with their own kind but also with their close relatives. The sight and sound of thousands of calling geese flying overhead against a dawn sky are awe-inspiring, and a colourful mixed flock of wild ducks on a winter lake is not only a beautiful sight but an exciting identification challenge for keen birdwatchers.

Classifying wildfowl

The group of birds we call 'wildfowl' are the ducks, geese and swans. They form the bird family Anatidae and are all somewhat similar to look at, in superficial ways at least, with broadly similar behaviours and lifestyles. They are all closely related, descending from a shared ancestor that lived some 25–35 million years ago.

The birds of the world are divided into 40 taxonomic groupings, called orders. Each of these broad divisions is further divided into one or more families. In the case of Anseriformes, the order to which ducks and geese belong, there are just three families. Worldwide, ducks and geese between them number 174 species, and along with about seven species of swans, they make up the family Anatidae. This family is by far the largest of the three in the order Anseriformes; the other two are Anhimidae (the screamers – peculiar chicken-like birds of South America) and Anseranatidae (a single species – the Magpie Goose, *Anseranas semipalmata*, a bizarre early offshoot of the lineage that evolved into modern ducks and geese).

Within each family there is at least one genus. The family Anatidae contains 53 genera. Each genus contains at least one species and often many more. For example,

Below: Its unusually long legs make the Magpie Goose (left) suited to a more terrestrial life than most other wildfowl. The Cape Barren Goose (*Cereopsis novaehollandiae*, right) is a peculiar species native to southern Australia.

Above left: The Spectacled Eider's close evolutionary relationship to the Common Eider is evident in its appearance.

Above right: Similar to other sawbill ducks, the Hooded Merganser of North America has a prominently crested head.

Below: King Eiders are rare visitors to the British Isles, but they form large flocks in Arctic waters.

the genus *Anser*, the 'grey geese', contains eight species; and the genus *Anas*, the dabbling ducks, includes about 30.

Every species has a scientific name made up of two words. The first word is its genus name, which it shares will all other species in its genus, and the second is its species name. So, for example, the Greylag Goose is *Anser anser*, while the Pink-footed Goose is *Anser brachyrhynchus*. The Mallard is *Anas platyrhynchos*, and the Pintail is *Anas acuta*.

New names, please!

Although scientific names are universal, they are not unchangeable. When scientific evidence comes to light that we have made a mistake with our classifications, a species may be moved from one genus to another, or into an entirely new genus of its own. With the advent of genetic research, many such discoveries have come to light. In 2009, a team of researchers at Heidelberg University in Germany looked in detail at the genes of the ducks classified in the genus *Anas*. Their findings showed that several of those species should be placed in new genera. For example, the paper recommended that the Shoveler and its close relatives did not belong in *Anas*, and should be classed separately in the genus *Spatula*. The British Ornithologists' Union, which maintains the official British List of birds, changed the Shoveler's name on its list from *Anas clypeata* to *Spatula clypeata* in January 2018 (these things take a while to work through the system!).

Above: The Shoveler – formerly *Anas clypeata* – is now called *Spatula clypeata*.

It is a source of frustration to birdwatchers that scientific names sometimes change, and that this makes otherwise perfectly good bird books out of date. But scientific names must reflect the current state of scientific knowledge. English names are much less likely to change. For example, some of the ducks in the genus *Anas* are called 'teals', but so are a range of other ducks from different genera.

The value of scientific names is twofold. First, they are understood universally. Birds all have their own local names in the languages of the countries where they occur, but scientific names are the same everywhere in the world. For example, the bird a British birdwatcher calls an Eider is called a *Honkewatagamo* by a Japanese birder, an *Edredone Commune* by an Italian speaker, and a *Kajka Morská* by a birdwatcher from Slovakia. But say *Somateria mollissima* and all four of those birdwatchers will know at once which species you are talking about. The other useful feature of the scientific name is that it includes the bird's genus name. So if you hear talk of an unfamiliar bird called *Somateria fischeri*, you will know at once that it is closely related to *Somateria mollissima*, and that will tell you it is a kind of eider (it is, in fact, the Spectacled Eider, found in Alaska and north-eastern Siberia). This, in turn, will give you an idea of what it might look like and what sort of lifestyle it might lead.

Ducks and geese – and shelducks

The family Anatidae is often split into several subfamilies, of which the largest are Anserinae and Anatinae. Anserinae includes the swans and the 'true geese'. Most of the other subfamilies contain species that we classify as ducks. The subfamily Anatinae is home to most of them, while Aythyinae contains the freshwater diving ducks.

The only other subfamily with representatives in the UK is Tadorninae. This subfamily contains the shelducks and their relatives, which are somewhat intermediate between ducks and geese. In the UK, we have just two members of Tadorninae – the Shelduck, *Tadorna tadorna*, and the non-native Egyptian Goose, *Alopochen aegyptiaca* (their common names reflect our confusion over whether they are ducks, geese or something in between!). To keep things simple, we shall consider these two species as ducks rather than geese in this book.

Tadorninae aside, it is usually easy to tell whether you are looking at a duck, a 'true goose' or a swan, at least in the UK. Swans are very large, very long-necked, elegant and usually pure white (though two species are black and white). Geese are also quite large and long-necked, but are smaller and stockier than swans, and are usually grey or brown, often with some black and white markings. Both swans and geese have relatively small heads and bills, and males and females look almost exactly the same. Ducks are the smallest and are short-necked, with proportionately larger heads than geese. Among ducks, the sexes usually look different, with males being colourful or boldly patterned, and females drab grey-brown. Our two species from Tadorninae are the only ones to throw a spanner in the works – they are colourful, but they are larger than the ducks and have rather goose-like body shapes, and the sexes look more or less the same.

Above: The Shelduck is our biggest and one of our most strikingly marked duck species.

Ducks and geese in the UK

Eight species of true geese and 22 species of ducks are well established in UK waters (inland and coastal) and can be reliably seen at the right places for at least part of the year, even if their numbers are small. Nearly all of our species are migratory to some extent, and their numbers increase in winter as breeding birds from more northerly countries – such as Iceland, Greenland, Russia and Scandinavia – arrive to take advantage of our relatively mild winters. Some of those that breed in the UK do so in only tiny numbers but become quite common in winter thanks to these influxes, while others are winter visitors only. One species (the Garganey, *Spatula querquedula*) is with us only in summer – it heads to Africa for the northern hemisphere winter.

A few of our wildfowl species are not here naturally but have established self-sustaining breeding populations after being introduced (deliberately or accidentally) by people. A couple more have some native and some introduced populations. Introduced populations are not typically migratory – in these cases, their winter

Below: The Teal is our smallest duck and, despite being rather shy, it is common in the British Isles in winter.

population comprises the breeding population plus their offspring from the year before.

The table on pages 14–15 lists all of the regularly occurring UK species of ducks and geese, with their populations (breeding and wintering) and their status in the UK (native, introduced or both). Population numbers are the most recent available, and they are taken from data supplied by the British Trust for Ornithology (BTO) and the Royal Society for the Protection of Birds (RSPB) – see page 125.

Above: Dabbling ducks, such as the Pintail (left), are typically longer-bodied and more elegant than diving ducks, such as the Tufted Duck (right).

Below: Uniquely among British wildfowl, the Garganey is a summer visitor to our shores.

Species	Breeding population	Wintering population	Native or introduced?
GEESE			
Canada Goose *Branta canadensis*	62,000 pairs	160,000 birds	Introduced (also occasionally occurs naturally as a wanderer from its native North America)
Brent Goose *Branta bernicla*	None	103,000 birds	Native
Barnacle Goose *Branta leucopsis*	900 pairs	99,000 birds	Both (native as a winter visitor, but breeding populations are introduced)
Greylag Goose *Anser anser*	46,000 pairs	231,000 birds	Both (native as a winter visitor, but most breeding populations are introduced)
Pink-footed Goose *Anser brachyrhynchus*	None	510,000 birds	Native
Taiga Bean Goose *Anser fabalis*	None	230 birds	Native
Tundra Bean Goose *Anser serrirostris*	None	300 birds	Native
White-fronted Goose *Anser albifrons*	None	14,100 birds	Native

Species	Breeding population	Wintering population	Native or introduced?
DUCKS AND SHELDUCKS			
Egyptian Goose *Alopochen aegyptiaca*	1,100 pairs	5,600 birds	Introduced
Shelduck *Tadorna tadorna*	15,000 pairs	47,000 birds	Native
Mallard *Anas platyrhynchos*	61,000–146,000 pairs	670,000 birds	Native (although numbers have increased through releases and escapes of farmyard and pet ducks)
Pintail *Anas acuta*	9–33 pairs	20,000 birds	Native
Teal *Anas crecca*	2,100 pairs	430,000 birds	Native
Wigeon *Mareca penelope*	400 pairs	440,000 birds	Native

Species	Breeding population	Wintering population	Native or introduced?
Gadwall *Mareca strepera*	1,200 pairs	31,000 birds	Native (although numbers have increased through releases of captive birds)
Shoveler *Spatula clypeata*	700 pairs	19,000 birds	Native
Garganey *Spatula querquedula*	14–93 pairs	None	Native
Tufted Duck *Aythya fuligula*	16,000–19,000 pairs	130,000 birds	Native
Pochard *Aythya ferina*	681 pairs	23,000 birds	Native
Scaup *Aythya marila*	1–2 pairs	3,900 birds	Native
Red-crested Pochard *Netta rufina*	10–21 pairs	570 birds	Introduced (also occasionally occurs naturally as a wanderer from mainland Europe)
Mandarin *Aix galericulata*	2,300 pairs	13,000 birds	Introduced
Goosander *Mergus merganser*	3,100–3,800 pairs	15,000 birds	Native
Red-breasted Merganser *Mergus serrator*	2,800 pairs	10,000 birds	Native
Smew *Mergellus albellus*	None	130 birds	Native
Long-tailed Duck *Clangula hyemalis*	None	13,000 birds	Native
Eider *Somateria mollissima*	26,000 pairs	81,600 birds	Native
Common Scoter *Melanitta nigra*	52 pairs	130,000 birds	Native
Velvet Scoter *Melanitta fusca*	None	3,400 birds	Native
Goldeneye *Bucephala clangula*	200 pairs	19,000 birds	Native

Above: The Mandarin is one of several non-native species to have become established in the wild in the British Isles.

In addition to these 30 'regular' UK species, there are several other species of ducks and geese on the official British List of birds. These 'extras' fall into two categories. The first are those species that stray here unpredictably, in small or very small numbers, from their native lands. The second are those that have been seen in the wild from time to time in the UK but are not thought to have reached our shores under their own steam. This second category of birds – escapees from captivity – may eventually be classed as established UK breeding birds if enough of them get together and breed successfully to form an enduring self-sustaining population. This has already happened with the Canada Goose, Egyptian Goose, Red-crested Pochard and Mandarin.

The full British List of birds includes three species of swans, 11 species of geese, and 40 species of ducks and shelducks, all of which are deemed either to occur (or have occurred) here naturally or to be self-sustaining introduced species. Another 47 wildfowl species (including four swan species) have been recorded in the UK as escapees from captivity – some of these are known to have bred in the wild too, but are not (yet) considered to be fully established and self-sustaining.

The 'plastic problem'

Every once in a while, a birdwatcher out and about in the UK will see something entirely unexpected. It might be a Budgerigar (*Melopsittacus undulatus*) flying with a flock of House Sparrows (*Passer domesticus*), or a Harris's Hawk (*Parabuteo unicinctus*) from America sitting on someone's garden fence. Many different kinds of birds are kept in captivity in the UK, and once in a while, they escape. These escapees do not tend to survive very long, and it is usually very obvious that they are indeed escapees (or 'plastic' in birdwatcher jargon) rather than wild birds that flew here from wherever they live naturally.

When it comes to wildfowl, though, things get more tricky. Many species of wildfowl in the northern hemisphere are strong flyers and migratory by nature, so are more than capable of turning up thousands of miles from home. As long as they find suitable habitat, they can live happily in their new environment for months or years. However, many of these wildfowl species are very popular as pets and are found in public and private aviaries and bird collections in the UK. They are usually kept in the open and prevented from flying away by wing-clipping, where the primary flight feathers are cut short on one wing. But new feathers grow each year, and if the keepers are not on the ball, the birds can regain the power of flight and escape.

Below: You might see some exotic ducks apparently wild in the British Isles, such as the Ringed Teal from South America (left); these ducks are certainly escapees from captivity. Escapee Bar-headed Geese, as seen here (right), have bred in the wild in the British Isles, although so far only in minimal numbers.

Above: North American migratory species, such as the Wood Duck, may occasionally arrive under their own steam, but the vast majority of these ducks will be escapees.

Below: The striking Muscovy Duck is popular as a backyard pet, and unwanted birds are often abandoned in town parks.

So, how can you tell whether the Hooded Merganser (*Lophodytes cucullatus*) you have found is a truly wild bird that has flown here from North America or one that has simply jumped over the fence at the local zoo? Escapees are often suspiciously unafraid of people, and any bird that turns up outside the normal migration season is also probably an escapee. Sometimes the matter can be solved definitively if the bird is wearing a leg ring. The ring's details will reveal whether the bird was ringed as a wild bird in its native country, or (much more likely) as a captive bird. However, in many cases, it is simply not possible to know for sure, and each record of these tricky birds has to be assessed on its own merits and a 'best guess' made as to its provenance. At the time of writing, only seven of the hundreds of British records of Hooded Merganser have been accepted by the British Birds Rarities Committee as being genuinely wild birds.

The 'alien' ducks and geese that are most frequently seen at large in the UK tend to be striking or colourful species that are popular in aviaries and ornamental wildfowl collections, such as the Wood Duck (*Aix sponsa*), the Ringed Teal (*Callonetta leucophrys*) and the Bar-headed Goose (*Anser indicus*). It is also common to

Duck-a-likes

Not every bird that swims and dives in the water is a duck. There are several other unrelated birds that have the same general 'look' as a duck on the water – a heavy boat-like body and strong swimming feet. Among them are the grebes, the divers, the Coot (*Fulica atra*) and Moorhen (*Gallinula chloropus*), and the auks. However, all of these birds have different bills to ducks – theirs are not flattened on the horizontal plane, and they end with a sharp, forward-aimed point rather than being blunt and rounded. There are other differences too, but you can always tell with a look at the bill whether your bird is duck or non-duck.

Above: Although it swims and dives with ease, the Coot's pointed bill and lobed toes indicate that it is not closely related to the ducks.

see escapee Muscovy Ducks (*Cairina moschata*) and Swan Geese (*Anser cygnoides*), as these two species have been fully domesticated for their eggs and meat, and in the UK they are often kept as backyard pets.

Returning to our 30 regularly occurring British geese and ducks, they can be categorised according to their appearance and general way of life into the seven following groups:

Grey geese are generally large, with grey-brown plumage. The feathers have pale edges, giving a scaly appearance, particularly on the back and flanks. The bill and legs are usually pink or orange. Most British species are winter visitors, and they frequent farmland and grassland. They belong to the genus *Anser*.

Taiga Bean Goose

Black geese range from duck-sized to close to swan-sized, and have grey, black and white plumage in a bold pattern. The bill and legs are black. Two species are winter visitors to estuaries and coastal grazing marshes, while Canada Geese are non-native and occur year-round at parks with lakes and on grassland. They belong to the genus *Branta*.

Brent Goose

Shelduck

Shelducks are larger and longer-necked than other ducks, and are colourful, with males and females looking almost the same. In the UK, we have only two species – the Shelduck (genus *Tadorna*), found largely in estuarine areas; and the Egyptian Goose (genus *Alopochen*), mostly found in parkland with lakes.

Dabbling ducks live mainly in freshwater habitats. They feed by grazing, dabbling and upending but seldom dive. Males are often very colourful, while females are drab. They are usually found in well-vegetated lowland waters, but also occur on larger reservoirs and coastal lagoons. In the UK, our dabbling ducks are in the genera *Anas*, *Mareca*, *Spatula* and *Aix*.

Male Gadwall

Freshwater diving ducks live mainly in freshwater areas. They primarily feed by diving underwater and swimming (propelled by their feet) to the bottom of the water bodies to take vegetation. They are not as colourful as dabbling ducks, but males are more boldly marked than females. They occur on deep lakes, and the Scaup is mainly seen offshore. In the UK, our freshwater diving ducks are in the genera *Aythya*, *Netta* and *Bucephala*.

Male Scaup

Subadult male Red-breasted Merganser

Sawbills are ducks with long, slim, serrated, hook-tipped bills. They feed mainly on fish, which they pursue underwater. Males are more colourful and boldly marked than females. They breed on upland rivers and winter on large lowland lakes. In the UK, our sawbills belong to the genera *Mergus* and *Mergellus*.

Seaducks live on the open sea in winter and dive deep to forage, mostly feeding on molluscs, which they pull from the seabed and crush in their strong bills. Males are usually black or black and white, while females are browner. Scoters nest inland on wet moorland, while Eiders nest on rocky seashores. Our seaducks belong to the genera *Melanitta*, *Somateria* and *Clangula*.

Male Long-tailed Duck

Evolution and Adaptation

When you are out birdwatching in winter, you may see ten or more species of ducks and geese at the same lake. Observe for a while, and you will notice more and more differences in the way they behave. Some paddle around in the shallows, while others (especially geese) seem to prefer to graze on the shores. Some feed in slightly deeper water, upending their bodies to reach the lake bed, while the diving ducks feed in the deepest parts. Some ducks sit in full view in large, noisy flocks, while others are shyer and more solitary, lurking in clumps of reeds and sedges.

Birds have evolved many different ways to feed in and on water. Because all birds still have to lay eggs and incubate them on dry land, they can never live on or over water full time, although many species have entirely severed their link with the land apart from these bare essentials. Outside the breeding season, birds like albatrosses and shearwaters live far out at sea – the water offers all the food they need, as well as a safe place to rest. They nest on small islands and are always just a few metres from the open sea. Other waterbirds, like most species of gulls, will feed on and around water at times and regularly swim on the surface, but in both summer and winter, they spend most of their time on the land.

Most ducks and geese fall in between these two extremes. They will feed and rest on the land at least some of the time, but their short, sturdy legs and fully webbed feet are more adapted to swimming than walking, and they are more comfortable (and safer) on the water. However, they are thought to have evolved from land birds. Their close relatives, the screamers, live in wetland areas but rarely swim – they have long, chicken-like legs and barely any webbing between the toes. The ancestors of Anseriformes were more similar to screamers than to modern wildfowl.

Opposite: Mandarins are normally found on lakes and slow-flowing rivers. However, they also possess strong sharp claws that allow them to perch on tree branches, which can be useful because they nest in tree-holes.

Prehistory

Birds that were recognisably ducks and geese appear much further back in the fossil record than most other contemporary birds. The fossil ancestors of wildfowl include birds in the family Presbyornithidae, which were found over most of the world not long after the end of the Cretaceous period, 66 million years ago, when dinosaurs became extinct (except for those lineages that were already on the way to evolving into modern birds).

One of the birds classified in Presbyornithidae was the peculiar-looking *Presbyornis pervetus*, probably the best-known prehistoric member of the order Anseriformes. This large bird had the build of a stork or other long-legged, long-necked wading bird, but its bill was broad, flattened and duck-like. It lived in Asia and America somewhere between 62 million and 55 million years ago. *Telmabates antiquus* was a similar-looking species present in South America at around the same time, while another member of Presbyornithidae, *Wilaru prideauxi*, was still living in Australia as recently as 22 million years ago, several million years after the first true wildfowl had begun to appear.

Below: Fossil remains of several individuals of *Presbyornis pervetus* – an extinct wading bird that is regarded as one of the first Anseriformes.

Hawaiian oddities

The extraordinary moa-nalo are among the avian lineages that did not survive to the present-day. These goose-like ducks evolved on the Hawaiian Islands but were wiped out when humans arrived sometime after AD 124. They were enormous, heavy, flightless birds with stout, robust bills and legs, and lived on land, grazing and browsing vegetation. No mammals (except bats) lived on the islands at that time, so the moa-nalo filled the same ecological role as deer and antelopes in other parts of the world. When humans arrived (and along with them, pigs, goats and other grazers), the moa-nalo were soon hunted and outcompeted to extinction.

Another bizarre extinct Hawaiian duck was the Kaua'i Mole Duck (*Talpanus lippa*), which lived to at least 4050 BC. Its skull and structure suggest it was likely both blind and flightless, and lived by foraging in leaf litter with its bill, probably walking about turning over the vegetation and using smell and touch to find its prey. Ecologically, it would have been equivalent to mammals such as hedgehogs.

Despite their peculiarities, both the moa-nalo and Kaua'i Mole Duck are true members of the family Anatidae and close relatives of modern ducks. They show how evolution, through natural selection, can remodel a familiar body plan in entirely unexpected ways if there are opportunities to exploit.

Right: An artist's impressions of the impressive moa-nalo, a duck relative that once lived on the Hawaiian islands.

The earliest fossils of birds considered to be Anatidae (the family comprising modern ducks, geese and swans; see page 7) include *Cygnopterus* from China (about 33–38 million years ago). This was a gigantic, tall bird, initially thought to be a relative of the flamingos because of its graceful long neck and legs, but is now considered to be a swan-like relative of the true wildfowl. The first fossil remains of what might be regarded as 'true ducks' are about 23 million years old and include *Manuherikia minuta*, a tiny diving duck from New Zealand. Fossil ducks of the recent genus *Anas* date back at least 15 million years, and similar-aged fossils of birds from the modern goose genus *Branta* have also been found.

Evolution

Understanding the basics of how evolution works means appreciating two simple things. The first is that all individuals in a population have a slightly different genetic make-up. New genes come into a population through random genetic mutations, and these can then be passed on to an animal's offspring. This genetic variation makes all the animals in a population slightly different physically and, in some ways, behaviourally. Individual birds in a flock will vary in their size, wing length, intensity of colour and pitch of voice, in the exact compass direction that they head when they begin to migrate, and in countless other ways.

The second key point is that the individuals most likely to survive, breed and pass on their particular set of genes are those best suited to their environment. Those less well adapted are more likely to die before they can breed. They might be too slow to escape a predator or have plumage too conspicuous to stay hidden, or they may migrate in the wrong direction and end up lost over the sea. This 'weeding out' of the less survivable individuals in a population is called natural selection. The result is a population that is, as a whole, better adapted to its environment. The population may also split into two distinct groups, both of which are well adapted but in different ways. This is how one species can become two – a process called speciation.

Right: Although superficially similar to other wildfowl, such as the Mallard, the Eider is adapted to a very different way of life. The former is a freshwater dabbler, and the latter is a deep-diving seabird.

Left: Across its range in North America, the Canada Goose has evolved into several distinct subspecies, which vary in size; smaller forms occur further north.

Left: Greylag Geese from East Asia (subspecies *rubirostris*) are bigger than those that breed in northern Europe and north-western Russia (subspecies *anser*).

If an environment remains more or less unchanged for millennia, the organisms living in it are likely to stay unchanged too. But when the environment changes more quickly, old ways of life may not work well any more, while new opportunities can open up. Any animals carrying a new mutation that benefits them in the changed environment stand a better chance of surviving and breeding. Their beneficial traits will spread through the population from generation to generation, and evolutionary change happens.

Sometimes, evolution simply cannot keep up with the rate of environmental change. Extinction, rather than adaptation and speciation, becomes more and more likely

Above: Modern birds have evolved to exploit our rich wetland habitats in many different ways, and to share these habitats in (relative) harmony.

as the rate of environmental change increases. Throughout history, several mass extinctions have occurred when a particular environmental change took place over a short space of time (relatively speaking) and caused catastrophic die-offs across all kinds of living things. The most recent of these occurred about 66 million years ago, probably as the result of a massive asteroid strike in the region of Mexico. Its impact would have caused a vast dust and soot cloud to form. This would have blocked sunlight worldwide for many months, causing an abrupt fall in sea temperature, widespread acid rain and other ecological problems. The disaster killed off about 75 per cent of all the world's animals and plants, including all of the larger land vertebrates – among them many early birds and mammals, as well as the dinosaurs.

As planetary conditions settled, those species that survived the event thrived, spread and diversified at an unusually rapid rate, refilling all of that biological space. Most modern bird and mammal lineages appeared over the 20 million years that followed the event, evolving and adapting to adopt lifestyles that were formerly the preserve of the dinosaurs. The ancestors of modern ducks and geese were no exceptions. They evolved from landbirds to waterbirds, and from there into dabblers and divers, grazers, snail-crushers and fish-catchers, and spread throughout the world.

From one species to two

Evolution is a dynamic process, and we can see it going on around us (albeit at what seems a very unhurried pace to us, with our brief lifespans). When a species contains distinctly and consistently different populations, we define each of these as a subspecies. This can be seen as an intermediate step along the process of speciation.

Among the ducks and, especially, the geese in the UK, there are several examples of subspecies. The White-fronted Goose is a large grey goose with a white 'blaze' around the base of its bill. It breeds in the Arctic and has five subspecies, two of which migrate here in winter: the pink-billed *Anser albifrons albifrons*, which breeds in north-eastern Europe and Russia; and the orange-billed *A. a. flavirostris*, which breeds in Greenland. The *flavirostris* geese tend to come to northern and western Britain, and *albifrons* to the south and east, but there is some overlap. Not only do the two subspecies look different, but they also have different behaviours. For example, *flavirostris* White-fronts associate with their offspring for an unusually long time, sometimes even taking care of their grandchildren, while *albifrons* parents do not. The two forms may rub shoulders in winter, but their breeding grounds are thousands of kilometres apart, suggesting that differences between them are likely to increase over time as they adapt more and more to their respective breeding environments. Some ornithologists consider that they are already dissimilar enough to be classed as distinct species, rather than just subspecies.

Above: White-fronted Geese of the pink-billed subspecies *albifrons* have prominent white fringes to the back feathers, giving a scaly appearance, and a broad white tail surround.

Above: The orange-billed subspecies *flavirostris* has a less scaly-looking back and more restricted white around the tail base and tip.

Physiology and anatomy

Right: Membranous webbing between the front three toes, as seen in this Mallard, makes the duck's foot a serviceable swimming paddle.

Ducks and geese have several key traits that are adaptations to life on the water, the most obvious of which is their foot structure. Like most birds, they have four toes, with the front three pointing forward and the rear backwards. The rear toe is small and does not usually touch the ground when the bird is standing up. In all ducks and geese, the front three toes are connected by membranes (webs), and in some diving ducks, such as scoters, the hind toe has a fleshy lobe. The webs (and, if present, the lobe) provide resistance when pulled through the water, allowing the bird to propel itself forward on the backstroke.

Right: Common Scoters and their relatives have a fleshy lobe on their hind toe for a stronger swimming stroke.

On the forward stroke, the toes are pressed together to cut down resistance. The leg and toe bones are also flattened vertically to present a narrower profile to the water on the forward stroke. Geese, and ducks that spend most of their time on flat ground when not swimming, tend to have broad and blunt claws, but ducks that spend time up in trees, such as the Mandarin, have sharper and more hooked claws to provide grip on branches.

Wildfowl also have a distinctive bill shape, although there is a lot of minor variation. The Shoveler has the most exaggerated 'duckbill' – its spoon of a bill is long and very flat, expanding to a wide, rounded tip. Geese have relatively short and strong, wedge-shaped bills, adapted to grab, cut and pull vegetation and to dig into the topsoil. In the fish-catching sawbill ducks, the bill is very narrow along its whole length and has a small but definite hook at its tip. Seaducks, which pull up and crush shellfish, have large, thick-based bills.

The inner edges of the bill are lined with comb-like projections called lamellae. Their function and structure vary between species. In geese, they are solid bumps that help provide grip when the bird is cropping and pulling up grasses and other plants, while in dabbling ducks they are longer and finer, acting as a sieve within the bill to strain small edible items from the water. Sawbills' lamellae are modified into sharp, tooth-like points, which help them get a grip on their slippery fish prey.

Below: The bill shapes of different species are adapted to the way they feed. For example, Greylag Geese (left) are grass-croppers, while Goosanders (right) are fish-grabbers.

Ducks and geese are heavy for their size and have relatively short wings that are broad-based and pointed. The wingbeats are driven by their large breast muscles. This makes them fast and powerful in the air – the Eider is capable of faster straight-line flight than almost any other bird on Earth. But their flight is not highly energy-efficient, and they cannot easily glide or soar like larger-winged, lighter-bodied birds; larger geese also need a good run-up to become airborne. When they want to land, geese perform a manoeuvre known as 'whiffling', where they turn sideways – and sometimes almost upside down. This enables them to lose height quickly while still keeping their wings spread and ready to flap, to control the final descent and awkward landing.

The plumage of ducks and geese is dense and insulating, with a thick down layer formed by individual fluffy down feathers, and also the fluffy bases of the larger, smooth contour feathers. Female ducks and geese shed some of the down feathers on their bellies when they are sitting on their eggs, to provide a soft lining for the nest and to expose warm bare skin for more efficient incubation. The flight feathers have strong shafts to withstand the forces passing through them when the bird is flying, with

Below: A Greylag Goose (accompanied by a second Greylag Goose and a Canada Goose) 'whiffles' to lose height in the air before landing.

Left: The dense plumage of diving ducks, such as the Tufted Duck, helps keep water away from their skin as they swim underwater.

flexible tips to help them make small speed and direction adjustments on the wing.

Feather colour is usually produced by melanin pigments within the feather cells. However, bright iridescent colours, such as the blue and green of a male Mallard's head, are the result of a specialised feather structure (which interferes with and reflects specific light colours) rather than pigment. Female ducks, although mainly drably plumaged, often have bold or bright wing

Below: Most female ducks have rather nondescript plumage that acts as camouflage for the many hours spent on or near the nest when incubating eggs.

Right: Iridescent plumage, such as the male Mallard's head, is not colourfully pigmented, but has a microstructure that reflects different shades of bright light – usually in the green/blue/violet range – depending on the viewing angle.

markings that show in flight – these probably act as signals to help keep a migrating flock together. Feather condition, flexibility and waterproofing are maintained through the application of preen oil secreted from the bird's uropygial gland, at the upper base of the tail. Ducks and geese are also very keen bathers – you may see geese, in particular, performing remarkable acrobatics on the water as they engage in lengthy bathing and preening sessions.

Below: Wildfowl, such as the male Gadwall, spend much of their time preening to keep their feathers in good condition.

Various other anatomical traits are related to life on and in water. Wildfowl are among only a handful of bird families in which the males have a penis; this makes it easier for them to copulate while moving on water. The respiratory system (lungs and linked air sacs) are well developed in ducks and geese, especially in diving ducks, which carry a larger volume of air in their respiratory systems to enable them to make long dives. All ducks and geese also have salt glands, structures in the head that extract excess salt from the blood (functioning like extra kidneys). The concentrated salty solution they produce is excreted from above the bill. The salt glands are particularly well developed in the seaducks, enabling them to drink the seawater on which they swim without ill effect.

Above: The vigorous splashing wing movements of this drake Pintail as it bathes help to dislodge any dirt from all parts of the plumage.

Breeding

With their bright eyes and fluffy coats, ducklings and goslings are about as adorable as baby birds can be. And as wildfowl families tend to be on the large side, you will occasionally see ten or more young in a close-knit brood. Duck and goose courtship is also charming to watch – for the most part – and parental behaviour, although varied, is impressive. These birds, more suited to swimming than walking or perching, must still take to the land to nest. Compared to their terrestrial cousins, they have a particularly challenging time when it comes to keeping eggs and young safe.

Swans are famously monogamous and usually stay with their chosen mate for life, but it is not only swans that could be used to symbolise lifelong love. The same goes for most species of geese, with pairs sticking together even through their long migratory journeys. Most ducks, however, are quite different. They do form short-lived pair bonds, but the male does not participate much (if at all) in parental care, and in many cases, he leaves his mate as soon as she is ready to lay eggs.

Opposite: Most baby wildfowl, like this Mallard, have drab-coloured, camouflaged down.

Left: When goose families rest, the female usually stays very close to the young while the male stands guard nearby.

Courtship

The one-parent-family approach adopted by ducks explains why the males in most duck species are highly colourful, while the females are drab. Males do not stick around while the female is incubating eggs, so do not require camouflaged plumage for those long hours spent sitting on or near the nest. Instead, they can afford to have colourful, ornamented plumage – all the better for catching a female's eye. Because a female duck does not need anything from her partner except a sperm donation, she needs only look at a male's physical condition to assess his worthiness as a mate. Duck courtship is therefore often a communal affair, in which males display their health and beauty with noisy and frequently comically contortionist displays, while females assess them and make a choice.

The display behaviours of male ducks usually involve postures that show off particularly prominent and contrasting plumage features. Many species lift their heads up and raise their breasts out of the water, sometimes also simultaneously cocking up their tail. Male Goldeneyes tilt their heads back onto their backs, while Pintails and Long-tailed Ducks bob and quiver their elongated tail feathers.

Right: Single parenthood for a female Mallard begins as soon as she has laid her eggs.

Left: In their communal courtship displays, groups of male Wigeons posture and call as they circle a female.

These movements are executed while the bird turns on the spot or swims rapidly forward in a short surge, often calling as he moves. Courtship calls are usually very distinctive and vary from the low, almost suggestive-sounding croon of the Eider to the sharp, high whistle of the Wigeon. Females often perform similar, though less exaggerated, versions of the male's movements, especially when a pair is together without other males around.

Left: The bizarre postures struck by an excited male Goldeneye help draw the female's attention to his physical fitness and the quality of his plumage.

Above: As the female Canada Goose incubates the eggs, the male stays at her side, ready to see off any threats.

Geese go about things very differently. They share all parental responsibilities, more or less. The female does the incubation, but the male stays close by to protect her, so both need plumage that conceals them reasonably well on land. When the goslings arrive, both parents will shepherd and defend them. However, the male is particularly vigorous in his defence, leading the family wherever they go and chasing after any perceived threats, while the female follows behind and remains closely attentive to the goslings. Goose courtship is a more even-handed affair than duck courtship, as both sexes pay attention to the other's behaviour. Females stick close to their mates, while males try to show off how strong and scary they are. In spring, you might see Canada Geese swimming about in pairs. The male of a pair (a little larger than the female) often attacks other males in a loud, aggressive display. This gives the female an idea of how willing he might be to take on a threat to their future goslings and how intimidating he might be when in attack mode.

Hybridisation

It is relatively unusual for animals of different species – even those that are closely related – to breed together. Incompatibilities at all sorts of levels, from different numbers of chromosomes to different courtship behaviours, help to prevent it from happening. When a mixed pair does breed (usually because one of the birds is unable to find a mate of its own species), eggs may not hatch, or if hybrid young are born, they usually have diminished or no fertility and sometimes other health problems as well.

However, hybridisation in ducks and geese is relatively common, even between species that are not all that closely related. The Mallard, for example, has been recorded hybridising with a total of 59 other species, including nearly all dabbling ducks of its own genus (*Anas*), but also members of 14 other genera, including both true goose genera (*Anser* and *Branta*). The popularity of mixed ornamental wildfowl collections does greatly increase both the opportunity for mixed pairing and for the results to be observed. However, hybridisation is also frequent in the wild. Hybrid Canada × Greylag geese are probably the bird hybrid most often seen in the wild in most of the UK.

If you find a strange duck or goose that you cannot match to any species in the books (including possible exotic escapees), it may well be a hybrid. Working out parentage can be very difficult, although in some cases (particularly in male ducks) hybrids show obvious traits from both parent species.

Above: A hybrid Canada × Greylag Goose; this is a frequent hybrid in town parks and around reservoirs in lowland UK.

Above: This male duck's distinctive auburn quiff indicates Red-crested Pochard parentage. The black-and-white body plumage reveals the other parent – a Tufted Duck.

Above: Hybrid Pochard × Ferruginous Ducks would, if seen in the UK, have probably originated from a captive collection of exotic ducks.

Mating

Ducks and geese may copulate on land or in the water. Mating usually follows a period of mutual courtship display of some kind (often synchronised head-dipping) and closeness between the pair. The female signals readiness to mate by lowering her head, after which the male mounts her back while gripping the feathers of her neck or the back of her head with his bill. This can look quite alarming when the pair is on water, as the female may be completely submerged for a few seconds at a time.

Wildfowl are among the few birds in which the male has an everted sexual organ – a penis. In most birds, males have no penis and copulation occurs via a 'cloacal kiss'. Both birds have a cloaca – a single opening through which droppings are expelled, and that also has a sexual function, passing on sperm in males and receiving sperm

Below: Mating looks an uncomfortable experience for female dabbling ducks, such as this Gadwall, as she would not usually immerse herself fully in the water.

and laying eggs in females. Mating involves the male bird bringing his cloaca to touch the female's and ejaculating sperm at the moment of contact. For ducks and geese, however, having a means for the male to penetrate the female's body before ejaculating makes it easier for the pair to mate while on water.

Battle of the sexes

Having a penis means that male ducks and geese can engage in forced matings with uncooperative females. Forced mating is most often witnessed among Mallards in spring. Once the females begin to lay eggs, their male partners leave them and go in search of other females, targeting any that have not yet started to nest. Often several males will 'gang up' on the same female, driving away her own mate if she has one and trying to mount her. The males are forceful and single-minded, and the resultant struggle can be violent and very distressing to watch. The female can end up losing a lot of feathers from her head and neck as the males grip and pull

at her, and occasionally some individuals even drown. Every so often, hormonally charged male Mallards will try to mate with females of other species, and there is even a record of one trying to copulate with a dead male Mallard.

A female Mallard may not be able to escape from a determined male, but her internal anatomy is complex, with extra openings and blind alleys that will trap sperm and prevent fertilisation. What we do not know for sure (although it seems likely) is whether the female can actively control what goes on after penetration, and thus effectively choose which male fathers her young.

Above: Forced group matings can be very dangerous for a female Mallard, but it's believed she does have some choice over which male (if any) will fertilise her eggs.

Nesting

Female ducks and geese build rather untidy nests from nearby vegetation, lined generously with soft feathers shed from their bellies. This feather loss also creates a brood patch – an area of bare skin with a rich blood supply that is in contact with the eggs during incubation. Most ducks and geese are rather hefty and not very agile on land, and they also have chicks that need to move around and find their own food very soon after hatching. For these reasons, they usually build their nests at ground level rather than up a tree or in some other elevated position. Nests on the ground are, of course, pretty vulnerable to predators, especially predators of the furry kind. If a Fox (*Vulpes vulpes*) were to find a duck on her nest, it would surely eat the eggs and the duck, too, if she did not escape quickly enough. The birds do, however, have a few weapons at their disposal.

For a start, ducks and geese choose well-hidden nest sites. They often place their nests within dense vegetation or under low, overhanging branches. Islands in lakes are also good nesting spots as they offer more safety from ground predators. Occasionally, Mallards nest in high

Below: A Tufted Duck is relatively inconspicuous in her nest.

Above: A female Goldeneye regards the world from her tree-hole nest. Soon, her ducklings will follow her lead and jump to the ground.

places such as on roof terraces, which keeps them safe from predators while they are incubating their eggs. However, this can cause new problems when the ducklings hatch and need to be shepherded safely to water – even a small, lightweight duckling may be injured from a long fall onto concrete. Mandarins and Goldeneyes nest in holes in trees, as do Goosanders – their ducklings are adapted to climb out of the hole and to survive the jump to the woodland floor or into the water below them. Natural tree-holes large enough for ducks are quite rare in the UK (partly because we do not have any very large woodpecker species), but hole-nesting ducks are quite willing to use purpose-built nest boxes. Egyptian Geese also often nest in trees, using natural hollows that form when a large branch breaks off and the wood decays around the breaking point. Shelducks are also sometimes hole-nesters, although not in trees – for them, an old Rabbit (*Oryctolagus cuniculus*) burrow is the usual choice.

Below: This brood of newly hatched Canada Geese will spend only a short time keeping warm in their nest before heading out into the world.

Female ducks of most species that breed in the UK are drab, mottled and brown, to help them blend in with the undergrowth, earth or leaf litter when they are sitting on

Left: The problem with tree-nesting is the leap to the ground – a scary prospect for Mandarin ducklings that are just a few hours old.

their nests. It is possible that their natural smell changes during the breeding season to become less detectable by mammalian noses – this has been noted in some other types of ground-nesting birds. When flushed from their nests, Eiders and some other duck species will defecate over their eggs as they leave, producing a particularly smelly kind of poo that repels mammals and so may keep the eggs safe.

Geese, being so much larger than ducks, are less able to hide their nests. The female goose incubates the eggs, but the male stays close and is ready to attack any intruder. You may have experienced such an attack, when a male goose approaches you with neck outstretched, hissing menacingly. The chances are that he has an

Above: Male geese, such as these Canada Geese, guard their mates closely during courtship, and this protective behaviour will only get stronger when there are goslings to care for.

incubating mate or a brood of goslings nearby, although this protective behaviour is often apparent even before nesting has begun. The gander's size and aggression are enough to discourage most of the mammalian predators that are likely to be a problem in the UK. Some species of geese also nest in loose colonies and use the force of numbers to drive away any danger.

Right: Most wildfowl lay pale beige, greyish or greenish eggs.

The incubation period for ducks and geese varies from about 21 days to 30 days, and does not begin until the female has laid all of her eggs. This is because it is imperative that all the eggs hatch at the same time, so the family can quickly vacate the nest area and reach the relative safety of the water. Many species will lay 10 or more eggs in a clutch, so a female could be laying eggs for well over a week (usually one every day) before she begins to incubate. The eggs of most species have dull-coloured shells to help provide some camouflage and the shells often become stained by damp vegetation over time.

Out of synch

In the northern hemisphere, nearly all birds begin their breeding activity in early to mid-spring, and their chicks hatch when spring is in full flow, and food supplies are at their most abundant. Baby birds need lots of protein, and goslings of even the most vegetarian geese will consume plenty of insects.

The Egyptian Goose, native to sub-Saharan Africa, is adapted not to the cycle of spring, summer, autumn and winter, but to the rainy/dry season pattern of the tropics. It naturally breeds at the end of the dry season, which in much of its range is in late winter. Egyptian Geese in Britain often have small goslings as early as February, which is not optimal for their survival (both because of potentially low temperatures and a lack of food and because there is no 'glut' of baby birds around to divert the predators' interest). Nevertheless, the species does breed here successfully, especially in more urban areas where predation is less of a problem and food supplies more reliable.

Above: It is not unusual to see Egyptian Geese with small goslings as early as February in Britain.

Hatching and early life

When ducklings and goslings hatch, they are initially weak and tired from the process of breaking out of the shell and need to rest for at least a couple of hours. Hatching can take the best part of a day. As soon as the first hole is made in the shell, the chick inside starts adapting to breathe in the open air and its body goes through some final changes, including drawing in its yolk sac, the contents of which will be its first food as a hatchling. When it is ready and strong enough, it enlarges the hole and forces its way out by pushing its feet against the shell.

After they have rested and their fluff has dried, ducklings and goslings can walk and swim, find food and feed themselves. They instinctively follow their mother, the first large moving object they see (a process called 'imprinting' – ducklings and goslings hatched in captivity in an incubator will just as readily imprint on their human carer), and once the whole brood is ready she leads them away from the nest. They will not return there, although they may still need their mother to help keep them warm, especially at night – they sleep on land, huddling under their mother. The young birds also need their mother to defend them from danger and to

Below: Ducklings and goslings hatch with damp down, but this dries off and becomes fluffy and body-warming very quickly.

Egg-dumping

Once in a while, you will see a much bigger brood of ducklings than usual. And occasionally you will encounter a clutch that has an interloper in its midst – an individual that is clearly a different species to the rest of the youngsters. How does this happen? It is well known that Cuckoos never rear their own young but lay their eggs in other birds' nests instead. This same behaviour (brood parasitism) is practised more casually by many kinds of ducks and geese.

When a female lays an egg in a nest that is not her own, she has no further investment in its future and so expends no energy in its care, but there is a chance the owner of the nest will accept this new egg as her own. Egg-dumping is, therefore, an effective way to increase the likelihood of rearing more young. Even if a female loses her own nest completely, she may still have young that year if she managed to dump an egg or two elsewhere. She will usually choose nests of her own species, but egg-dumping in the nest of another species is not unusual. Baby birds reared by another species may seek a mate of that species when they mature, which could result in hybridisation.

lead them to suitable places to feed. In the case of geese, the father is always present too, and the family typically travels in a line, with the father leading and the mother at the rear.

Goslings spend a lot of time on land, feeding on grasses, while ducklings feed mainly in the water. The ducklings of dabbling ducks are more enthusiastic divers than their parents and find much of their food in this way; they will also dive to escape danger from above.

Below: This female Red-crested Pochard lost all but one of her ducklings when they were small. The sole survivor is now big enough to be too much for many predators to handle.

Crèches

Some species of wildfowl that habitually nest in quite close proximity will combine their broods to form large crèches. Sometimes this can happen accidentally, such as when two broods get muddled up and all the ducklings or goslings end up following the same parents, but in other cases the formation of crèches is habitual. The species best known for this is the Eider, which tends to nest on rocky shores and islands in loose colonies. An Eider crèche may contain several dozen ducklings, tended by a few, often unrelated females that did not breed that year. Crèches may help to reduce predation through the presence of extra 'guards'. In the case of Eiders, the ducklings feed mainly on the surface while adults find more suitable food for themselves by deep diving. So, crèching behaviour may actually be of more benefit to mother Eiders than to their ducklings, freeing them up to visit the most productive feeding areas and recover from the rigours of egg-laying and incubation while a team of 'aunties' take care of the chicks.

Above: A mixed-age crèche of Eider ducklings with their 'auntie'.

Ducklings especially are very vulnerable to all kinds of predators, and even the most protective mother is unlikely to keep them all alive. In the UK, birds such as harriers, gulls and herons can easily catch ducklings, as can large predatory fish like Pike (*Esox lucius*). On land, mammals like Foxes and Stoats (*Mustela erminea*) are problematic, and Otters (*Lutra lutra*) and the non-native American Mink (*Neovison vison*) can take ducklings just as easily on land as in the water. Many females will lose their entire brood, although as numbers dwindle it is easier for them to protect those that do survive.

Above: Gulls are regular predators of Eider nests, though females do their best to keep them away.

Fledging and independence

Ducks and geese are born independent in terms of feeding themselves, and as they grow larger and replace their fluff with feathers, they no longer need to be brooded at night. After a few weeks they are at, or close to, adult size, are able to fly, and no longer need to be protected from predators (or at least they are no more vulnerable than their parents). Nevertheless, young geese will carry on associating with their parents all through their first winter. When you see flocks of wintering geese, you can often pick out the family groups within the flock – this is particularly true of species such as White-fronted Goose, where the first-winter birds look different to adults (they lack the white face marking).

Most ducks do not associate closely with their parents after they become independent. Instead, both adults and youngsters form flocks with others of their species, and they will spend the rest of autumn and winter living socially, often alongside gatherings of other species. The young birds' first (juvenile) plumage is usually very similar to that of adult females, but they moult from juvenile to adult plumage through late autumn and into winter, with

Below: The juvenile plumage of the White-fronted Goose (left) lacks the adult's charismatic white face and blackish belly barring (right).

the young males becoming almost identical to adult males. Among most ducks, young birds of both sexes will engage in courtship behaviour and form pairs in their first winter, ready to breed in spring, while Shelducks breed at two years old. In geese, breeding does not usually occur until the age of three.

Above: White-fronted Geese have developed extensive blackish markings on their bellies by the time they reach breeding age.

Left: Although still in juvenile plumage, this young Shelduck has full-grown flight feathers and will be living independently.

Diet and Feeding

On a visit to a park, many of us will have been amused by the sight of a Mallard dabbling, and from time to time tipping forward in the water and pointing its rear to the sky. If Tufted Ducks are also around and the water is clear enough, you might see them dive and then swim rapidly underwater, chasing drifting food morsels. The local geese, meanwhile, are probably up on the bank, cropping away at the grass as efficiently as a herd of sheep. Despite the similarities in their bill shapes, ducks and geese have a considerable variety of feeding styles and dietary preferences.

Feeding behaviour

The type of food a bird eats can often be ascertained by aspects of its appearance, particularly its bill shape. Birds of prey have hooked bills to tear at flesh, while seed-eating birds usually have robust, deep-based bills and strong jaws to crack shells. Long-billed waders like Snipe (*Gallinago gallinago*) and Curlew (*Numenius arquata*) find worms and other invertebrates in soft soil, while birds that catch flying insects have wide, bristle-lined mouths to trap their prey. The standard 'duck bill' has a flattened shape and is lined inside with comb-like lamellae (see page 31). This

Opposite: 'Upending' enables these Mallards to reach food on the lake bed.

Left: Wigeons are often seen grazing on damp meadows in winter.

Murderous Mallards

Many ducks are omnivores and opportunistic. A few observers have seen Mallards catching, killing and eating fish, and even on occasion fledgling songbirds. They are physically ill-equipped to perform this task but seem to undertake it with enthusiasm nevertheless. Mallards at a reservoir in Romania were observed flushing prey from the undergrowth and then catching it on the water. A fledgling Black Redstart (*Phoenicurus ochruros*) and a Grey Wagtail (*Motacilla cinerea*) were both caught, drowned and then vigorously shaken before being swallowed. New feeding behaviour like this, if successful, can soon be learned by other members of the flock. For example, Mallards in coastal California have learned to catch crabs in the sea, a behaviour first documented in 2013 but now observed at multiple locations.

Above: A Mallard duckling makes an unlikely meal of a dragonfly.

shape works well to sieve small items of food out of water, but it is also good enough for grabbing and pulling at soft vegetation. A few small evolutionary tweaks to this bill shape in different wildfowl lineages have modified it into an effective cutting, gripping or crushing tool.

Dabbling ducks

When dabbling ducks dabble, they are taking water into their open bills and then pushing it out again with the bill closed. Water passes through the gaps in the lamellae, and all but the tiniest solid items are kept within the bill. Larger items – which may include insects and other small animals, as well as bits of plants – can be grabbed and swallowed as they are. Most dabbling ducks, and especially the big-billed Shoveler, spend a lot of time feeding on the surface in this way. They will also submerge their heads to reach underwater, and upend to reach deeper still, where they can grab and break off bits of vegetation growing underwater. They occasionally dive, but this is more likely as a tactic to escape a predator than for everyday foraging. The Wigeon feeds

on land more frequently than other dabbling ducks, often cropping grass in the manner of a goose. The Gadwall is a dabbler but also a part-time thief. Gadwalls stay close to Coots when the latter are busily diving to find the underwater vegetation they eat, and they will try to grab the food from a Coot's bill as it surfaces.

Above: This foraging Coot has attracted the close attention of a pair of would-be thief Gadwalls.

Left: Teals mostly find their food by dabbling in shallow muddy water.

Freshwater diving ducks

The freshwater diving ducks, such as Tufted Ducks and Pochards, have a similar diet to the dabblers but theirs includes a greater proportion of animal life, including aquatic insects and molluscs like water snails. They can dabble, but they mainly forage underwater. When swimming below the surface, they are outstretched and sleek, and move quickly with fast twists and turns, using strong strokes of their webbed feet to propel them along. When actively feeding, Tufted Ducks usually dive for about 20 seconds at a time, or sometimes a little longer, with pauses on the surface of about 10 seconds before the next dive. They usually swim down to a metre or so but can easily reach depths of 2m (6.5ft) or more during a long dive.

Right: Diving ducks, such as the Tufted Duck, swim quickly and with great agility underwater.

Shelducks

Shelducks feed most often on muddy estuarine shores, and their diet is overwhelmingly dominated by a single species of snail, the Mudsnail (*Hydrobia ulvae*), which they sift out of the soft surface mud. They also take other invertebrates and a small amount of vegetation. The Egyptian Goose, a close relative of the Shelduck, takes mainly plant matter.

The stance of a duck on the water can reveal whether it is a diver or a dabbler. Dabbling ducks look more buoyant and their rear ends, in particular, stand proud of the water's surface. Their tails are usually relatively long and pointed. Tufted Ducks and Pochards have a lower profile, and their tail tips sit only just above the surface. Some deeper-diving species, such as the sawbills and seaducks, sit even lower in the water and often swim with the tail tip underwater. Males of the dabbling Pintail and the deep-diving Long-tailed Duck both have very long central tail feathers, and while the Pintail's tail feathers are invariably in plain view as the bird swims along, the male Long-tailed Duck's feathers are often not as they lie flat on the water's surface.

Swimming posture also indicates the diving habits of other waterbirds besides ducks. Swans and geese do not dive and their swimming posture is high-backed, with upward-pointing tails. The same goes for gulls and Moorhens. However, grebes and divers have a flatter, lower

Above: The upturned rear of a male Pintail can be easily distinguished from other ducks by its long, tapering tail.

profile and their rear ends barely clear the water, while Cormorants (*Phalacrocorax carbo*) and Shags (*P. aristotelis*) sit lowest of all, with their long tails entirely submerged.

Above: Shelducks 'dabble' in mud to find their favourite mollusc prey.

Sawbills

In the sawbill ducks, the lamellae in the bill are separated into pointy, tooth-like tips. These give the bill a serrated edge that provides grip on fish. In these birds, we therefore see something perfectly analogous to teeth – the lamellae have the same function as the little pointed teeth of fish-eating mammals such as river dolphins. But sawbill 'teeth' are made from lightweight keratin (like the rest of the bill) rather than heavy enamel and dentine – birds jettisoned real teeth early in their evolutionary history, when the development of flight imposed a selective pressure for their bodies to be as light as possible. Sawbills are underwater hunters that swim quickly and with agility in pursuit of fish. When hunting, they will swim on the surface with their heads down to watch the water below, diving under with a slight jump when they spot prey. A Goosander can stay underwater for nearly a minute while chasing fish.

Right: Fish are hard to hold onto and many fish-eating birds, including the Goosander, have bills adapted to pierce and grip the slimy skin of their prey.

Seaducks

The seaducks are all deep divers, swimming down to the bottom of the sea to find their food, which comprises mainly mussels, crustaceans, sea urchins and other invertebrates that dwell on the seabed. An Eider can swim underwater for well over a minute, as can scoters, and they can reach depths of 30m (100ft) or more. Uniquely, the Long-tailed Duck uses its wings as well as its feet to help propel itself underwater, enabling it to reach greater depths than any other diving duck – up to 55m (180ft).

Eiders feed mainly on mussels, pulling the bivalves from their moorings on the seabed and then carrying them to the surface in their bill. They also take crabs and sea urchins. Smaller items may be swallowed whole, but others require some processing – clumps of mussels need to be shaken apart and large mussels need to be crushed, sea urchins are rolled in the bill to snap off their spines, and crabs are shaken forcefully to remove their legs before the body is swallowed. Seaducks take different food in the breeding season; this is especially true of scoters, which nest inland on moorland lakes and at this time feed mainly on insects.

Adult geese are almost entirely vegetarian and eat a great deal of grass and other growing plants, as well as seeds and grains. The hard lamellae in their bills act as scissors, snipping through plant stems, and the birds may also root in the topsoil for tubers. Some Brent Geese wintering in the UK consume eelgrass, a plant that grows in muddy estuaries that is in decline. The Pink-footed Geese that come to us in winter primarily forage on farmland, feeding on cereal stubble, grassland, winter-sown cereals and the crowns of Sugar Beets (*Beta vulgaris*), which are discarded during harvesting. Goslings require more protein in their diets than adults, and eat quite a number of small invertebrates.

Below: Our wintering Brent Geese feed on eelgrass found growing on estuarine mudflats.

Feeding the ducks

There has been conflicting advice in the press about whether it is alright to feed the ducks, geese and swans at your local pond, and if so, what they should be fed. Bread is the usual choice, but it is not nutritionally optimal (this is especially true of white bread); this is terrible news for parkland wildfowl, some of which may eat little else but bread. There is some evidence that nutritional deficiencies cause the condition 'angel-wing', in which the wrist joints in the wings of ducklings and goslings are misaligned, to the extent that the flight feathers stick out at an odd angle when the bird reaches adulthood and it cannot fly. It is undoubtedly the case that lots of uneaten bread in the water can cause algal blooms to develop, which harm all aquatic life. However, the habit of feeding bread is so established at some lakes that it has supported larger populations of wildfowl than would naturally be present. In 2018, there were several news reports of how sudden withdrawal of bread feeding was causing birds to starve in certain areas. In urban areas particularly, we can expect to lose a lot of parkland wildfowl if everyone stops feeding the ducks.

Above: 'Angel-wing', as seen with this Egyptian Goose, is a developmental deformity that may be diet-related.

An occasional slice of bread will do little harm, but avoid feeding if there is already uneaten bread lying around in the water. Floating pellet foods made especially for wildfowl are available, and regular birdseed that you might feed to your garden birds is quite acceptable too. If you have a large garden with lawns and some water, you may well have visiting Mallards. These ducks are quick to work out that an easy meal can be had underneath a bird table or bird feeder.

Above: Feeding the birds in your local park is a great way to observe them at close quarters, but it's important to offer the right types of food.

Digestion

Unlike most birds, ducks and geese do not have a crop. This outpouching of the oesophagus allows some storage of food before digestion proper begins. In ducks and geese, the whole oesophagus is stretchy and expandable, so can be packed with food that can be digested later at the bird's leisure. This allows the bird to consume a large volume of food rapidly – a useful trait, as ducks and geese are often disturbed by predators while feeding and have to make a quick escape.

From the oesophagus, the food reaches the first part of the stomach proper – the proventriculus, where it is softened and digestive enzymes begin to break it down. The next stage sees the food entering the second part of the stomach – the ventriculus, or gizzard. This is a very muscular organ that effectively 'chews' the food for the bird, breaking it up with powerful contractions. Ducks and geese swallow a certain amount of hard, indigestible material (small stones and grit) that stays in the gizzard and assists with the food-grinding process. Like other muscles, the gizzard can grow larger and stronger through use, but it will shrink if the bird's diet consists entirely of soft material.

Some protein absorption occurs in the gizzard, and further absorption of nutrients and water takes place as the food passes through the small intestine. Here, enzymes secreted from the liver (via the gallbladder) and the pancreas are added to the food to aid breakdown of nutrients. The food then reaches the large intestine, part of which is a two-horned structure called the caecum, a pouch containing bacteria that breaks down plant material through fermentation. The rest of the large intestine mainly reabsorbs any excess water, before whatever is left is excreted through the cloaca.

Below: A Brent Goose's winter diet is almost entirely vegetarian.

Social Life

It is rare to see a duck or goose on its own. These birds are inherently social, at all times of year. A female duck may be alone during her incubation, but after that, she will be with her ducklings for several weeks, and then she and they will join the adult males, which will already have been flocking together for some time. Geese famously go around in large, close-knit flocks, feeding, migrating and resting together, and often nesting close to other pairs. Sociality does not necessarily always mean harmony, though. Even within a huge flock, each individual bird is following its own priorities.

Being social is common in birds, although it is by no means universal. Birds of prey usually prefer to hunt alone, and some songbirds are aggressively territorial, not tolerating the company of any others of their own species, even in winter. Some other birds, although not strictly territorial, have no particular social ties – these include some seabirds, which will gather at an abundant food source but are also willing to wander alone for days.

Food and safety are the main reasons why birds are social (or antisocial). A hawk or an owl needs to hunt down and catch each meal, and the presence of another raptor in the same area trying to catch the same prey would not help

Opposite: Barnacle Geese are very sociable and gather in flocks to feed on land.

Left: Mallards and other ducks may form very large flocks in winter.

– nor is there enough to share. However, supplies of the food targeted by ducks and geese are often abundant and widespread once the birds are in the right place – perhaps a field of grass or a large lake. Trying to monopolise a large food supply by driving off rivals is energy consuming – it is better to use that energy feeding rather than rushing around. Of course, there are limits to how large or dense a flock can be before all its members are struggling to find enough to eat, at which stage there will be hostilities and some birds will be forced to move on.

As long as there is enough food for all, the advantages of sociality to each individual outweigh the disadvantages. The presence of others can indicate where food is. And feeding alongside these birds means that many eyes are on the lookout and each individual does not need to be as vigilant as when it is on its own. It also means that each bird is individually less likely to be a victim if the predator attacks the flock – the chances are that there will be others in the flock that are slower, weaker and closer to the point of attack. It is also more difficult for a predator to select an individual target among a large group than it is to attack a lone bird. Staying with the flock for roosting can also be life-saving on cold nights, when body heat can be pooled and shared.

Below: When danger – in this case, a Marsh Harrier – threatens, duck flocks take off en masse. It is difficult for a predator to home in on one individual within the group.

Flocking

The tendency of ducks and geese to join others of their ilk is well known. Indeed, this behaviour has been much exploited by hunters over the years with the use of decoys and duck-call whistles to encourage overflying birds down to land.

Some species are inclined to form very large flocks, of a thousand or more birds. In the UK, winter flocks of Brent, Barnacle and Pink-footed Geese in particular can be very large, as can gatherings of Wigeons and Common Scoters. Others are rarely seen in large flocks, even where food and other resources are ample. This may be down to circumstance rather than an inherent trait of the species. Garganeys, which are summer visitors to the UK, are nearly always seen here in pairs or alone, but on their African wintering grounds they form very large flocks. Tundra Bean Geese are usually seen here alone, in pairs or in small groups, and often with other goose species (especially Pink-footed Geese), but at most, only a few hundred in total will overwinter in the UK. Most of the world population winters in mainland northern Europe, where flock sizes can reach into thousands of birds.

Below: Nearly half a million Wigeons winter in the UK, most of them gathering in large flocks.

Above: The classic V-shaped skein allows all of these Pink-footed Geese (except the leader) to save energy by 'slipstreaming' the bird in front.

Migration is associated with flock formation – birds benefit from migrating in groups through sharing knowledge of where to go. Flying in groups is more energy efficient than travelling alone, too – a real boon for heavy birds such as geese. As long as they fly fairly close together, each bird receives some additional free uplift from the wingbeats of the bird in front, which is why they fly in a 'V' formation or in lines, with the front position constantly rotated as the leader tires and drops back. Those ducks and geese that are resident in the UK – mainly non-native species such as Canada Geese, Egyptian Geese and Mandarins – are also gregarious but do not form such vast flocks as some of the migratory species.

Flock sex ratios

In winter, male ducks are looking their very best and stand out much more than the drabber females as you scan across a lake. You might think that this is why there appear to be more males than females, but in some cases there is a genuinely uneven split between the sexes. The Pochard in particular shows a very male-heavy sex ratio in the UK, and there are less strong (but still definite) biases among Shovelers, Tufted Ducks, Wigeons and other freshwater species.

Why this should be the case is not immediately obvious, as these species show the usual 50:50 split at

their breeding grounds. Observations in the UK have shown that winter flocks of Pochards are dominated by males in more northerly areas, with the ratio evening out somewhat further south. This suggests that, of the birds arriving from eastern Europe and Russia, females generally migrate further south than males to find suitable wintering grounds. There is evidence that males can tolerate slightly lower temperatures than females, but the most convincing reason for the difference is that males arrive earlier than females at wintering grounds, and where flock sizes outstrip food supply, it is the females that lose out. Female Pochards are, on average, slightly lighter than males and this, coupled with their higher minimum temperature tolerance and the fact that breeding takes more of a toll on their body condition than it does on males, makes them generally less able to fight their corner where competition exists. Missing out on closer and more convenient wintering grounds and being forced to undertake a longer winter migration is a deadly combination for some – female Pochards suffer significantly higher winter mortality than do males, dying primarily from starvation.

The opposite is true of our sawbill ducks, particularly Smews. These small sawbills breed in the Arctic and are winter visitors to the UK in small numbers. Most of those that we see here are female-like 'redheads', rather than the stunning white-plumaged adult males that are so

Above: Male Pochards dramatically outnumber females in the UK in winter.

Right: 'Redhead' Smews outnumber white adult males in the UK in winter; some are females, but others are juvenile males.

coveted by birdwatchers. The UK is the southern limit of the Smew's wintering range, and (as with other ducks) it is the females that migrate further south, so it makes sense that we would see more of them. However, juveniles of both sexes resemble adult females, and juvenile plumage is retained well into winter. So, the 'redhead' Smews we see will include some juvenile males, as well as adult and juvenile females.

Odd one out

The various ducks and geese that occur in the UK only as scarce visitors, rarities or vagrants are typically discovered by birdwatchers looking through flocks of commoner species. Pink-footed Geese winter here in large numbers and a patient check through their flocks could produce Tundra Bean Geese, and perhaps a stray Snow Goose (*Anser caerulescens*) from North America. The Red-breasted Goose (*Branta ruficollis*) from Siberia generally turns up with flocks of the dark-bellied subspecies of Brent Goose (*B. bernicla bernicla*). Among ducks, the Green-winged Teal (*Anas carolinensis*) is usually found among Common Teals and the American Wigeon (*Mareca americana*) among European Wigeons. Vagrant seaducks also regularly associate with their closest relatives – King Eiders (*Somateria spectabilis*) with Eiders, and Black Scoters (*Melanitta americana*) and White-winged Scoters (*Melanitta deglandi*) with Common Scoters.

Above: A male King Eider nicknamed 'Elvis' regularly joins Common Eiders on the Ythan estuary in Scotland.

For these 'lost' birds, the safety of the flock may be enough to keep them there for a lifetime and they never try to find their way 'home', even though this means their chances of breeding successfully are much reduced. For example, a male Black Scoter found within a Common Scoter flock off Llanfairfechan, Wales, was seen there every winter from 1999 to 2007.

Communication

Birds that fly and feed in flocks need to stay in touch with one another, and birds that are social need to have ways to send particular messages. These may be vocal or visual, or a combination of both. The most fundamental message is the contact call made by birds flying or feeding together, each just saying 'I am here' to its neighbours to ensure that the group stays together – this is particularly important when visibility is poor, such as when flying through fog. The nature of the call varies from species to species, from the tuneful bugling of Pink-footed Geese to the high-pitched *whee-oo* whistle of the male Wigeon. Among ducks, the sexes often give clearly different vocalisations and females tend to have lower-pitched calls. Males of many species have specialised, partly bony structures in their throat called bullae, which are part of the trachea and provide resonance to their calls, making these both louder and clearer.

When a flock needs to act together in a hurry, for example if a predator turns up, the first bird to notice the danger will alert the rest with an alarm call and then all the birds will take to the air. Flying together at high speed and with frequent changes of direction can confuse a bird of prey enough that it cannot make an attack, as it needs to single out an individual to strike. Smaller ducks such as Teal are particularly agile in the air. The near-synchronous

Left: The vivid speculum markings found in many ducks – as seen on this female Mallard – can help them recognise one another.

Above: Maintaining flock cohesion while flying fast is a challenge, but species-specific calls and markings help the birds keep track of each other.

swirls and turns made by the flock are thought to occur as each bird reacts to the smallest direction changes of its immediate neighbours, creating a rippling effect.

Visual signals can also help with flock coherence. Most ducks and geese will form mixed flocks in winter, but by preference associate most often with their own species. Markings like the coloured wing-patches (speculums) found on both sexes in many duck species help them to recognise each other.

In winter, communal courtship displays take place within duck flocks. This social activity is often triggered by a female approaching males on the water, with a flattened body posture and frequent head-bobbing. The males respond by circling her with heads and tails raised, and performing plumage-shaking actions. It is only later that the female and her chosen mate will conduct courtship behaviours away from the group.

Right: Courtship for Pintails and other dabbling ducks involves postural displays on the water and chases in the air.

Interspecies interactions

Some of the benefits of sociality apply whether a duck or goose is flocking with its own kind or with other species. This is apparent when vagrant species turn up in the UK – they are nearly always found with other closely related species (see box on page 72). Sometimes the vagrant will have migrated some distance with the 'carrier species' flock.

More common species will also flock and feed together, but they often make small behavioural changes to reduce competition for the same resources. Wigeons and geese feeding together will eat more quickly than when alone, so their net food intake remains the same. There is also evidence that ducks may imitate the foraging techniques of other, more abundant species as well as using their own methods, and obtain more food as a result.

Below: Although their feeding behaviour is rather different, Pintails and Pochards are quite happy to rest and fly together.

Migration

Many northern hemisphere ducks and geese breed most abundantly close to or within the Arctic Circle. Here, they can take advantage of the intense (though brief) Arctic summer, when there are abundant supplies of insects for ducklings and less pressure from predators. When the day length begins to shorten and the temperature plummets, they leave, flying south for more clement climes in which to pass the winter months. The spectacle of thousands of wild geese or ducks on the move is one of the most visible signs of the changing seasons and is a sight (and sound) to quicken the heart of any birdwatcher.

Migration is a common strategy among wild birds to make the most of opportunities available in different areas at different times of year. In temperate zones, mid-spring to mid-autumn is a time of warmth and abundant food supplies – especially for insect-eaters. This is the best time for birds to breed. As autumn progresses, it becomes more difficult to find insect prey, so insect-eating birds must either switch to a different diet or migrate somewhere where their preferred food is available through winter. In the UK, thrushes and tits do the former, changing to a more vegetarian diet of fruit and seeds, while our swifts,

Opposite: Eiders are northerly breeders, but they wander around the whole UK coastline in winter.

Left: Wintering wildfowl come to the UK as early as August, but the main arrivals occur throughout September, October and November.

swallows, flycatchers and many warblers do the latter, migrating across the Mediterranean and the Sahara to spend their winters in Africa.

Not all northern hemisphere breeding birds have to travel as far to reach acceptable wintering grounds. The resident Blackbirds (*Turdus merula*) and Song Thrushes (*T. philomelos*) that stay in the UK in winter are joined by others that breed in continental Europe and Scandinavia, as well as Redwings (*T. iliacus*) and Fieldfares (*T. pilaris*). And while many of the Arctic-breeding ducks and geese that overwinter here also winter further south, few of them go much beyond southern Europe. Even so, this may still entail regular journeys of more than 3,000km (2,000 miles) between breeding grounds and the wintering site.

The summer duck

The Garganey is an oddity among British wildfowl in that it is only a summer visitor here. Its breeding range extends across Europe and through central and northern Asia to Japan. It is a long-distance migrant across its entire range, heading mainly to the southern hemisphere in winter. Birds in the UK and western Europe migrate to sub-Saharan Africa, while those breeding further east spend their winters in southern and south-eastern Asia and even as far as New Guinea. Garganeys arrive in the UK from March and have mostly departed by mid-October. They are very shy when nesting, so you are most likely to see them in April and early May, before breeding begins, and again in late August and September, once the young birds have fledged and when both adults and youngsters are preparing for migration.

Although it is closely related to the Shoveler, the Garganey also shares some similarities with two species of wading birds – the Ruff (*Calidris pugnax*) and Black-tailed Godwit (*Limosa limosa*). All three of these species breed in the same sort of lush, swampy meadowlands in the UK (and all are rare). They also all winter in similar parts

Above: Our experience of Garganeys in the UK is very different to that of people living in sub-Saharan Africa, where the ducks overwinter in large flocks.

of Africa, especially the Senegal Delta, where the Garganey is by far the most numerous duck species in winter. Most British Garganeys migrate south via France, Spain and the Straits of Gibraltar, but return via the central Mediterranean – a so-called 'loop migration', which helps them make the best use of the prevailing winds. Because Garganeys pair up while in their winter flocks and arrive with us when they are ready to breed, most of those we see in spring are twosomes, and it is unusual to see any larger gatherings.

Migratory behaviour

Birds 'know' when it is time to migrate through various factors, most importantly changes in daylight length, but also through having built up sufficient fat stores to fuel the trip, and changes in food availability. In spring, hormonal changes as birds near their breeding condition are also important. These cues work together to trigger behaviours related to migration, such as gathering in larger flocks and orientating toward the direction in which they will be flying. The urge to migrate and awareness of the correct direction are genetically determined, but young birds will follow the experienced elders within the flock on their first journey. In the case of geese, young birds associate closely with their own parents throughout migration. Over time and repeated migrations, they learn landmarks and other navigational cues.

Below: Wigeons on the move can be heard a long distance away, thanks to the males' high-pitched whistling calls.

Above: Barnacle Geese are true Arctic birds, but this very cold environment is too hostile for them in winter.

The migrant birds will depart when weather conditions are suitable. Often they will not make the full journey in one go but will stop at several 'staging posts' en route, to refuel. These sites are used year on year and can be as important to the survival of the migrating birds as the final wintering destination. However, some migratory routes involve crossing wide stretches of inhospitable habitat where no suitable staging posts are available.

Timing of migration is important to avoid food shortages and to escape bad weather. On the return journey to the breeding grounds, there is an extra time pressure – the need to arrive early enough to secure a good nesting site and to have time to breed, but not so early that food is scarce and the weather is bad. Those birds that arrive early will reap the rewards if local weather conditions are kind. Northward migration among geese is more predictable than that of some other birds, as it tracks the spring growth of plants. This means that the birds always arrive at staging posts and,

eventually, their breeding grounds just as new, nutrient-rich shoots are appearing.

Wildfowl vary in how strongly inclined they are to winter at the same site each year. Garganeys wintering in the Senegal Delta show no site loyalty to speak of, instead moving around according to the location of the most suitable feeding habitats. This flexibility is a useful trait for birds that winter in areas with unpredictable rainfall and flooding patterns. The geese that winter in the UK tend to be more site loyal. For example, one ringed Pink-footed Goose was sighted at Martin Mere in Lancashire for at least six successive years. At times, this individual also visited other sites in northern England and Scotland. Flocks of wintering geese will move around when necessary, as winter progresses and food supplies become depleted. Dark-bellied Brent Geese tend to spend more time feeding inland on grassland and cereal fields than on saltmarshes in mid-winter and early spring, returning to the saltmarshes later on as eelgrass and other estuarine plants begin to grow again.

Above: Dark-bellied Brent Geese feed in fields when eelgrass supplies run low in the estuaries.

MIGRATION

Moult migration

Some birds make a (usually short-distance) migration to a particular place where they will stay while they undergo their annual moult, at or near the end of the breeding season. They may then continue migrating, or return to where they started from. Having a safe place to moult is especially important for wildfowl because they are temporarily flightless at this time – unlike most other birds, which shed and regrow their flight feathers more gradually and retain the power of flight.

The Shelduck is a famous moult migrator, forming vast moulting flocks through July and into August. The majority of British adult breeding Shelducks leave their partly grown young in crèches under the care of just a few adults and set off for the moulting grounds. They mainly head for sheltered estuaries at the Wadden Sea, which straddles the borders of Denmark, Germany and the Netherlands, but there is also a significant moult gathering on the Mersey estuary in north-western England, holding some 20,000 birds. By the end of August, numbers are falling as the birds return to their breeding and wintering areas, clad in a new set of feathers and able to fly again. In some other duck species in which males do not participate in the care of the young, the males head to communal moulting areas once the females have begun to incubate their eggs.

Above: By early summer, Shelducks begin to moult, and many will soon migrate to their traditional moulting grounds to complete the process.

A changing world

Climate change, and its impact on the natural landscape, is having a distinct effect on the migration habits in wildfowl. Some Barnacle Geese have accelerated their return migration to the Arctic, to the detriment of their breeding success, though other populations are successfully adapting to the impact of climate change by utilising new staging posts and breeding further north.

A future with potentially earlier springs, warmer and drier summers, more prolonged winter cold and an increased risk of severe weather events will impact on wildfowl in many different ways. Migration is an adaptive habit, and perhaps wildfowl will be able to continue to adapt to the rapid changes occurring in the natural environment today. That said, many of the declines we are noting today among various species are at least partly attributable to climate factors.

Above: Climate change is affecting the migratory behaviour of species such as Barnacle Geese.

Left: The evolution of migration as a strategy has enabled birds to breed in areas that are only hospitable for part of the year.

Monitoring migration

Much of our knowledge of bird migration comes from marking individual birds, so that they can be identified again if recaptured (or found dead). Bird ringing (or 'banding', as it is known in some areas of the world) has revealed the migratory routes and destinations for a great many species, as well as allowing longevity, annual survival rates and other aspects of population biology, to be investigated. As optical equipment and camera technology have improved, researchers increasingly use marking methods that can be observed in the field without any need for the bird to be recaptured. These include leg rings with engraved characters that are readable in the field. Individually coded neck collars can be used in a similar way, for swans and some geese. Neck collars have the advantage over leg rings of being observable when the bird is on water, or standing in tall vegetation.

Today, we also have the option of tagging birds with electronic tracking equipment. This can enable us to trace the exact pathways taken by these individuals, revealing much more detail about exactly which way they fly, how long their journeys take, how long they spend at staging posts, and the effects of weather and other temporal events on migration habits. Using electronic tracking devices is typically much more costly, but can give far more detailed information from a single bird than a simple leg ring – different devices can provide fine detail of a bird's movements over weeks, months or even years.

Tracking devices include light-level geolocators – data loggers that record light levels relative to a fixed clock. This allows estimation of daylight length, and the time of solar midday each day. This information can be used to estimate the bird's position on the surface of the earth. These geolocators can be very small devices, although the accuracy of the locations they estimate is limited,

Below: Leg rings are used to individually mark both captive and wild birds, as seen here with this Barnacle Goose.

and typically the bird needs to be recaptured to retrieve the tag and its data. Then there are more sophisticated devices, which communicate the bird's exact position in real time, using satellite arrays, mobile phone networks or specially designed receiving stations. The accuracy of the locations recorded by these devices can be very high if GPS capability is included, but location accuracy and lifespan of the tag usually come with the costs of increased tag size and weight, and price. The choice of how to attach electronic devices to birds is also very important, and will differ depending on the size and ecology of the bird, and the size and capabilities of the device. In some situations, devices can be attached to a leg ring or neck collar, whereas larger tags may require some type of harness, like a small rucksack or climbing harness. Extensive research and checking should always be done before a combination of device type and attachment method are chosen, to consider bird welfare and ensure that bird behaviour is not significantly affected.

Most of what we now know about duck and goose migration comes from ringing and tracking studies. Ringing recoveries confirm that dark-bellied and light-bellied Brent Geese (the subspecies *Branta bernicla bernicla* and *B. b. hrota*) come to us from separate breeding areas

Above: Neck collars are easier to see than leg rings on swimming birds that spend a lot of time with their legs underwater, such as the Greylag Goose.

(Siberia, and Greenland, Svalbard and eastern Canada, respectively). Light-bellied Brent Geese may migrate 6,000km (3,750 miles) or more to reach our shores. We also know that the Barnacle Geese that winter here come from two separate areas (although unlike the Brent Geese, they are not separate subspecies). Those that winter in the Hebrides and western Ireland breed in Greenland, while those that winter around the Solway Firth in south-west Scotland breed in Svalbard.

Technology also plays a part in counting wildfowl populations, particularly on wintering grounds and at migratory staging posts. Aerial photography provides an easy way to count the numbers of Shelducks at their moulting-migration sites. Aerial surveys are also helpful for counting seaducks, which are difficult to see in detail from land.

Below: Birds feeding on grass, like these Brent Geese, tend to space themselves out and to move slowly, so are quite easy to count.

Although leg rings are the most common way to mark individual birds, various other methods are also used. You may see a goose wearing a neck collar, for example – a short plastic tube that fits around the neck and bears an engraved alphanumeric code. Wing tags, which fit on the leading edge of the wing, are also sometimes used. Data loggers and satellite trackers may be added to these kinds of tags, or fitted separately (see pages 84 and 85).

Neck collars, wing tags and coloured leg rings are used because of their visibility over a long distance. Welfare of the marked birds is the highest priority when designing and fitting the markers. Field observations show that the birds do not show signs of stress and don't attempt to remove the tags. Studies on tagged birds' long-term survival and breeding success also indicate that the tags do not impede natural behaviour or impact on survival rates. Catching and handling techniques used by ringers are designed to minimise stress, and newly ringed birds usually return to normal behaviour soon after release. The small costs of handling the birds have to be weighed against the benefits of tagging for the species as a whole. Tracking birds' movements helps us work out which areas are most important to them throughout the year. Using this data we can identify the most critical breeding and winter areas and migration staging posts, and then use that information to lobby for better protection of these areas. For example, satellite tracking of Lesser White-fronted Geese (*Anser erythropus*) breeding in Russia in 2018 revealed some hitherto unknown wintering sites for this endangered species in Uzbekistan. This finding subsequently led to a project to raise awareness of the species and its protected status, particularly with local hunters, and to safeguard the recently discovered sites.

Above: Fitting a leg ring necessitates handling the bird, which inevitably causes some stress. However, any discomfort needs to be weighed against the benefits and information gained by tagging the birds.

Threats and Conservation

Ducks and geese face the same dangers that affect most wild birds. Many will be killed by predators (often before reaching breeding age), and many more will die of starvation, while others may suffer death by disease or misadventure. The significant threats that face species as a whole are, unfortunately, nearly all down to human activity, whether directly or indirectly. Several wildfowl species around the world have become extinct over the last few hundred years and many others (including some British species) are endangered, although several species have recently been brought back from the edge of extinction by concerted conservation efforts.

One of the reasons ducks and geese produce such large broods is the very high risk of predation in the first weeks of the young birds' lives. Many of us will have seen a female Mallard at the local park, shepherding her large brood of tiny, newly hatched ducklings. A week later, however, the chances are that their numbers will have dwindled significantly. A duckling is an easy meal for many predators, with little ability to defend itself – it can run and

Opposite: Although it is not present in the UK, the Arctic Fox is a significant threat to wildfowl in some areas. It will patrol riverbanks and coastlines looking for birds and eggs. However, as seen here, if threatened female Eiders can ferociously defend their nest from predators.

Left: Most Mallard ducklings fall victim to some kind of predator in the first week of their life, especially if they venture away from their mother who is their main source of protection in their early days.

Right: The first week of life for a young duckling is when it is most vulnerable to predators. Here, a Velvet Scoter mother tries to rescue her duckling from a Herring Gull (*Larus argentatus*).

dive, but its main chance of escape is if its mother manages to fend off the danger. Her efforts, although invariably valiant, are hampered by the impossibility of keeping close to all the ducklings at the same time.

After the first week or so of life, the youngsters stand a much better chance, and not just because there are usually fewer of them by then, making them easier to defend. They grow rapidly and soon become too much of a mouthful for most predatory pond fish and opportunistic

Below: Goosander ducklings may save themselves from attack by predatory fish by riding on their mother's back.

Left: Foxes hunt adult ducks and ducklings alike. Geese are better able to defend themselves from most dangers.

urban Foxes, corvids and gulls. The majority of town parks do not have any other predators to speak of, but in the wider countryside there are other dangers, such as Marsh Harriers (*Circus aeruginosus*), Stoats, American Minks and Otters, all of which will attack bigger ducklings and even adult ducks.

Being larger, goslings are less vulnerable than ducklings – it also helps that they are guarded by both of their parents from the start. Many of our geese breed in the Arctic, where the main dangers to their nests and young are Arctic Foxes (*Vulpes lagopus*) and, sometimes, Polar Bears (*Ursus maritimus*). Predators of adult ducks and geese in the UK and elsewhere include Foxes and large, powerful birds of prey such as Peregrine Falcons (*Falco peregrinus*) and Golden Eagles (*Aquila chrysaetos*).

Adult wildfowl are not easy prey in general. On the water, they can dive to escape danger from above – this is second nature to diving ducks, of course, but dabbling ducks and even geese will also dive under these circumstances. In the air, they are fast and powerful flyers. They may easily be caught if weakened or injured, but sometimes birds with disabilities that prevent flight will survive for months or even years, especially in urban environments where predators are few and food is always available.

Wildfowling

Humans are also significant predators of many British wildfowl species. Shooting ducks and geese for food is a long-standing and, today, tightly regulated pastime in the UK. Although it is obviously still controversial, it is less contentious (from an environmental point of view) than the gamebird industry, which in the case of Pheasants (*Phasianus colchicus*) relies on the release of vast numbers of non-native birds in the countryside, and in the case of Red Grouse (*Lagopus lagopus*) is associated in some regions with the illegal persecution of birds of prey.

The legal quarry duck species in England, Wales and Scotland are Gadwall, Goldeneye, Mallard, Pintail, Pochard, Shoveler, Teal, Tufted Duck and Wigeon. It is also legal to hunt Canada, Greylag and Pink-footed Geese, and at present (and only in England and Wales), White-fronted Geese (although most shooting clubs operate a voluntary no-hunting moratorium for this declining species, which may receive full protection in the near future). In Northern Ireland, the Scaup is also on the quarry list. Hunting quarry wildfowl is permitted from 1 September to 31 January on inland waters, and 1 September to 20 February below the high-water mark on tidal waters (to 31 January in Northern Ireland).

Right: Most wildfowlers use dogs to retrieve shot birds from water.

Above: Wildfowlers are drawn to the challenge of shooting down fast-flying ducks, but shooting with a camera is an increasingly popular and no less challenging alternative.

Against the odds

As we saw in the 'Social Life' chapter (pages 66–75), ducks and geese feed and fly in large flocks for several reasons, including reducing their own odds of being lunch for a predator. If you want to lose yourself in the crowd, it really helps if the others in the group all look just like you. A bird within a flock that looks different is easier for a predator to track when the flock is on the move and is more likely to be the chosen target. This is why ducks and geese tend to form flocks of their own kind – even if they are mixed up together during feeding, they often separate into single-species groups when they take to the air. It is also why an oddity in the group – whether a lost individual of a different species or a bird with a plumage aberration such as leucism (white or partly white feathers) – has a survival disadvantage. Human hunters also tend to aim for the birds that stand out.

Above: A leucistic Greylag Goose stands out among its normal-coloured flock-mates.

Most wildfowl are shot on marshes owned or managed by shooting clubs, which expect their members to adhere to strict codes of conduct. The British Association for Shooting and Conservation gives advice on appropriate conduct. Shooters must be competent at bird identification so that non-quarry species are not killed, and they must bring a trained gundog to retrieve all shot birds from water. There are no official national limits on the number of birds a wildfowler may shoot, but clubs often impose their own limits – in practice, most wildfowlers will take only two or three birds per shoot. These factors help to ensure that wildfowling remains sustainable. Of course, even though only a few birds will die, shooting disturbs all the birds on a marshland. That disturbance may be harmful in itself, preventing natural behaviour – for example, studies show that some ducks may feed more at night than in the daytime on or close to shooting marshes. In spells of prolonged and severe winter weather, clubs may suspend all shooting to give the birds a chance to rest and feed in peace.

Uncontrolled hunting has, historically, caused problems for certain species in the UK and in other parts of the world. It continues to be seriously problematic in countries such as Iran, where large nets set up at night are used to catch whole flocks of migrating ducks.

Eiderdown

Besides shooting, another way humans exploit wild ducks is in the harvesting of eiderdown, which is a traditional practice in Iceland. However, this is done without any harm to the birds or their nests. The down is gathered from active nests and replaced with an alternative nest-lining material that works just as well. The down harvesters repay the birds by providing strict legal protection for them and their habitat.

Right: Eiderdown is a rare example of a natural resource that can be harvested entirely sustainably and without disturbance.

Pest control?

In addition to being a quarry species, the Canada Goose is also listed on the UK's general licence to control birds as a species that can be killed year-round by any landowner under certain conditions, without the need for special permission. The permissible reasons for doing this are: to prevent the spread of disease; to stop serious damage to livestock, food, crops, growing timber, fisheries or inland water; to preserve public health and safety; and to conserve other wild birds. The non-native Egyptian Goose is also listed on the general licence. It is expected that non-lethal methods of control should always be tried first and lethal methods used as a last resort, and anyone using the general licence must adhere to all animal welfare laws.

There have been calls from some people involved in angling and fisheries for the Goosander to be added to the general licence, on the grounds that the birds have increased considerably in number and are damaging fish stocks. While the species did increase between 1980 and 1995, it has since declined, and so far no change to the law has been made. Fisheries owners may apply for individual licences to cull Goosanders if they can demonstrate that there is a problem and that non-lethal methods have failed.

Above: Some fishery owners are lobbying for the fish-eating Goosander to be added to the general licence.

Disease and injury

Sociable birds like ducks and geese can be hit hard by outbreaks of infectious diseases. They are susceptible to various viral and bacterial diseases, and can also harbour internal and external parasites, which may weaken them and make them more vulnerable to predators. One of the most notorious diseases wildfowl can catch and spread is avian influenza ('bird flu', caused by the H5N1 virus and related strains), which can occasionally infect people and cause serious illness and even death. Because some ducks and geese are long-distance migrants and may visit town parklands and fields used by free-range poultry, they are of considerable concern during outbreaks of bird flu. Dead wildfowl found in the UK should be reported to the Department for Environment, Food and Rural Affairs (Defra; see page 125), and good hygiene should always be practised when feeding wild birds and handling dead or sick individuals.

Accidents that may befall wild ducks and geese include collisions with overhead wires, wind-turbine blades or other obstacles. Occasionally, they may become trapped in ice, or they may be hit by traffic. Geese are the most frequent species involved in birdstrike incidents and

Below: Wildfowl are fast but not especially agile fliers, which makes them vulnerable to collisions with obstacles, such as wind turbines, in poor visibility.

can cause serious damage to aircraft – if they strike the windshield or are sucked into the engine, the plane may be downed. In January 2009, a US Airways flight was famously brought down over New York by a birdstrike involving a flock of Canada Geese. The pilot managed to glide the plane to a ditched landing on the Hudson River, and all those on board survived.

Ducks and geese that are affected by injury or illness are much more likely to be taken by predators than are healthy birds, but they are also at high risk of starvation – an arguably more unpleasant death.

Above: The majority of birdstrike incidents, which can cause serious damage to aircraft, have involved geese.

Left: Road traffic is a hazard to ducks and geese living in urban areas.

Old age

Wild animals of all kinds rarely live out their potential lifespans. Only those few animals with no natural predators stand much chance of making it to old age, and as soon as their ability to find food is compromised, they are likely to starve.

We know from ringing recoveries that ducks and geese can be relatively long-lived. All of the geese that occur in the UK are known to be able to reach their late 20s, with one Pink-footed Goose reaching the amazing age of 40 years and 11 months. The longevity records for ducks are similarly impressive, with most species able to reach at least their early 20s. One Tufted Duck shot in Denmark was found to have been ringed 45 years and three months previously, and a UK-ringed Eider reached the age of 36 years and 10 months.

There are few signs of ageing in wild birds, but one that you may observe is the tendency of female ducks to acquire more male-like plumage with the passing years. These birds, known (inaccurately) as 'intersex', develop this unusual plumage as their female hormonal balance is altered by age. In areas such as city parks where there are few or no predators, it is not unusual to find an 'intersex' Mallard or two. They stand out particularly in the breeding season, looking distinctly different from normal males and females, but are less obvious in late summer and autumn, when male plumage becomes much more female-like ('eclipse' plumage). At all times, most intersex birds can be told from males by the variable blackish markings on their bills (true males have pure yellow or greenish-blue bills, even when in eclipse plumage).

Above: Female ducks can appear more male-like due to a hormonal balance that is altered with age. These birds can be distinguished from males by the blackish markings on their bill.

Wider concerns

The greatest danger affecting wildfowl and wildlife in general today is loss and degradation of their natural habitats – the places where they live and breed, and find all the resources that they need to be able to do so. This may take many forms. We tend to picture habitat loss as being the direct destruction of wild places to make way for buildings, farmland and other development. But habitats can also be damaged by more subtle factors. For example, water abstraction at one place could partly dry out a marshland elsewhere. Or if a non-native species is released in an area and becomes established, it could deprive native species of food, nest sites or other vital resources. And climate change and altered weather patterns can change the extent of regular seasonal flooding and droughts.

All of these human-caused changes, both overt and subtle, affect wildfowl populations. Not having enough good-quality breeding habitat will reduce breeding success, and poor-quality habitat in winter reduces survival rates as well as the breeding condition of those birds that do make it through the winter. The signs of these threats are not as obvious to us as, say, the remains of a bird killed by a predator. Instead, we notice the problems gradually – flock sizes slowly shrink, season by season, and numbers of breeding pairs diminish.

Left: Threats – including pollution, loss of food supplies and, in some areas, overhunting – have led to the Eider being considered Near Threatened on a global scale.

Right: The Velvet Scoter, an uncommon winter visitor to the UK, has suffered a dramatic population crash since the 1990s.

In the UK, many of our wildfowl species are now considered to be of conservation concern. This may be because of a small and beleaguered breeding population, or because numbers in winter have declined. The categories of conservation concern are Amber (for moderate threat) and Red (for significant threat); those species not currently of conservation concern are classed as Green.

When it comes to the risk of global extinction, we must consult the International Union for Conservation of Nature (IUCN). This body assesses the degree of conservation concern for wild species internationally, assigning each species to a category reflecting its current level of threat (Least Concern, Near Threatened, Vulnerable, Endangered and Critically Endangered).

While most of our regularly occurring duck and goose species (listed in the table on pages 14 and 15) are categorised as Least Concern on the IUCN Red List, the majority have Amber classification in the UK – only the Tufted Duck, Goosander and Red-breasted Merganser are on the Green List (not of national conservation concern). Six species – White-fronted Goose, Pochard, Scaup, Long-tailed Duck, and Common and Velvet Scoters – are on the UK's Red List and facing significant threat here. Of these Red-listed species, the Pochard, Long-tailed Duck and Velvet Scoter are also considered in danger on a global scale, and the Eider (Amber-listed in the UK) is currently categorised by the IUCN as Near Threatened.

Getting involved

As well as financially supporting nature conservation bodies, you can help tip the odds for wildfowl by taking part in surveys, and by volunteering at nature reserves. The British Trust for Ornithology's Wetland Birds Survey (WeBS) began life in 1947 under the name National Wildfowl Counts. Today, volunteer surveyors submit their counts of non-breeding waterbirds from some 2,800 wetland sites (both inland and coastal) each year, and the data generated are used to look at population sizes and trends (see Further Reading and Resources, page 125, for more information on surveys).

The RSPB and other organisations with nature reserves are always in need of volunteers to provide practical day-to-day help. The type of work varies from habitat management to

Above: Volunteers are vital contributors to our collection of knowledge about changes to wild bird populations. Here, a group of volunteers are carrying out a gull survey on the summit of Ailsa Craig, Firth of Clyde.

engaging with visitors, and you do not need to be an expert at wildlife identification. To find out more about helping at RSPB reserves, visit the RSPB website (page 125).

Of the three British duck species that are in danger of global extinction, two are seaducks. Many other North Atlantic seabirds are also struggling as a result of marine pollution, seabed damage, overharvesting of fish and other marine animals, and the effects of climate change on the distribution of prey. The Velvet Scoter, probably our most threatened wildfowl species and classed as Vulnerable by the IUCN, has suffered a decline in numbers of about 60 per cent since the 1990s and has a world population of just 300,000. In 2018, an International Single Species Action Plan for the Velvet Scoter was prepared by a collective of conservation bodies, detailing the threats faced by the species and ways to mitigate these. The threats named in the plan are wide-ranging, including loss of food supplies as a result of seabed trawling for scallops, overhunting, accidental killing as fisheries by-catch, sea pollution, predation by non-native mammals in the breeding season, and disturbance on breeding grounds.

The Pochard is a familiar species to most British birdwatchers and is easy to find on lakes and reservoirs in winter, so it may seem surprising that this bird is at risk of global extinction – its IUCN status is Vulnerable. However, our wintering Pochards are dwindling in number as a result of problems on their breeding grounds. The species has long been a rare breeding bird in the UK and western Europe, and while it is much more abundant in eastern Europe, its breeding population here has suffered a decline of 30–49 per cent since the 1990s. This explains the decline in our wintering population – a fall of 69 per cent between the early 1990s and 2015/16. The declines are probably mainly due to loss of breeding habitat and the effect of fertiliser run-off from farmland, which causes overgrowth of aquatic vegetation in the pools where the birds feed.

Based on counts of wintering birds in the Baltic Sea, the Long-tailed Duck – which is also classed as Vulnerable by the IUCN – underwent a severe decline in the late 1990s and early 2000s. It is often caught in gillnets (large nets that float near the surface, designed to trap fish by their gills) and is affected by oil and other pollution in many of its wintering areas. It is also heavily hunted in some regions.

The Eider has an IUCN classification of Near Threatened, meaning that it may be close to qualifying for a threatened category in the near future. Its moderate declines appear to be linked to low duckling survival because of insufficient food supplies. It is also affected by overhunting, pollution, entanglement in fishing nets of various kinds and overharvesting of the mussels on which it feeds.

Below: The Long-tailed Duck, one of our regular winter visitors, is considered at risk of extinction due to threats including habitat loss and oil pollution at sea.

Conservation

When population surveys reveal that a bird's population is in significant decline in a particular country, the government of the nation concerned needs to make a decision about what, if anything, to do about it. However, because most of our wildfowl are migratory, protecting their populations requires a cooperative multinational approach. Protecting breeding and wintering grounds will not mitigate the harm if there is uncontrolled hunting along a species' migratory flyway. Sometimes, there is indeed just one localised threat, but more often there is a range of problems to tackle, and countries vary significantly in how highly they prioritise conservation efforts and how much funding and expertise they have at their disposal.

In the UK, the RSPB is the main body concerned with conservation of birds and their habitats. The charity's work includes purchasing and managing land for the benefit of wildlife, advising the government and private individuals on matters related to bird conservation, and raising awareness among the public about these issues. At the time of writing, the RSPB manages some 130,000ha (320,000 acres) of countryside as nature reserves. Much of this land is marshland, estuary, wet meadowland and riverside – all habitats for ducks and geese. Other organisations such as the Wildlife Trusts and Wildfowl & Wetlands Trust (WWT) also manage land for the benefit of wildlife.

Targeted conservation plans for particular species depend on research, first and foremost, to pin down exactly how that species uses its habitat and the nature of the threats it faces. Then measures can be taken to help improve its situation. For example, the UK Biodiversity Action Plan for the Common Scoter, a rare and declining breeding bird of wet open moorland in the Scottish uplands, has so far initiated full surveys of the breeding population to assess ideal habitat conditions, and facilitated the removal of forestry plantations from key parts of actual and potential breeding habitat.

Wildfowl and People

Ducks and geese have long been important to humans the world over. They are frequently a source of food, and in temperate regions, their departure heralds the changing seasons. Their beauty is much admired, as is their prodigious reproductive capacity and, in the case of geese, their faithfulness. Some species have been domesticated for thousands of years and, thanks to their charming characters, are kept today as pets, as well as for their meat, feathers and eggs. Ducks and geese crop up in our sayings, from 'love a duck' to 'silly goose' and figure prominently in ancient legends as well as modern-day children's stories.

Myth and legend

Most ancient cultures have their own stories about how the world came into being, and often these creation myths involve animals of one kind or another. One of the old Egyptian stories sees the world created by a mighty goose, the Great Cackler, who laid an egg from which all of creation hatched. Another theme that appears in several stories casts a duck or goose as land builder, bringing mud up from the depths of a watery globe to create the solid continents. According to the indigenous Yokuts people in what is now California, the Sierra Nevada mountains and the California Coast Ranges were built by a duck after a great flood covered the whole nation with water. The duck did not do this on a whim, but was bribed to do so by an eagle and a crow, each of which wanted a mountain range to call their own.

The Egyptian Goose was a sacred animal in ancient Egypt, while domesticated Greylag Geese were kept

Opposite: Beatrix Potter's story of Jemima Puddle-Duck and the 'gentleman with sandy whiskers' teaches us a valuable lesson about the perils of naivety.

Below: Birds played a key role in the lives of ancient Egyptians, and Egyptian Geese in particular were considered sacred. This wall carving, dated at c. 2400 BC, showcases the significance of wildfowl in ancient Egyptian culture.

Above: This statue of Hans Christian Andersen, author of *The Ugly Duckling*, stands in Central Park, New York. As one of the most famous fictional ducks, the Ugly Duckling turned out to be a cygnet and not a duck after all.

as temple guardians in ancient Rome. The value of geese as 'guard dogs' is acknowledged today, and back in AD 390 it was geese at Rome's Temple of Juno Moneta that foiled a night raid by the Gauls on the city. The birds raised the alarm in typical goose style, rushing at the invaders with much hissing and flapping. Troops were soon on the scene to repel the Gauls, and thereafter the occasion was celebrated each year with a golden goose carried in a parade through the city. Today, geese rather than dogs stand guard at some Chinese police stations in Xinjiang province and have foiled at least one break-in, and they are also widely used in the province to guard flocks of chickens at private residences.

Copper goose idols, representing a goose god, were worshipped by the Finno-Ugric people of Scandinavia and Siberia a couple of thousand years ago. Prayers and sacrifices offered to the god would, it was believed, guarantee plenty of wild ducks and geese for the people to eat.

Ducks and geese have long been admired for their ability to straddle two worlds – they have mastery of both air and water. The Hindu god Brahma rode on the back

Right: Brahma, the Hindu god, portrayed on his white gander mount.

Love and devotion

If you get married in South Korea, you may be given a pair of carved ducks as a gift. This could be construed as a discouraging sign, given that ducks as a rule are not known for their lengthy liaisons. However, the carved ducks are intended to represent Mandarins, which are associated with fidelity and devotion. Although Mandarin males do, like other male ducks, leave the female in sole charge of incubation and raising the young, and do not necessarily pair up with the same bird again the following year, they do not appear to have the same promiscuous ways of many male ducks, and it is almost unknown for them to hybridise with other duck species. Nevertheless, geese would make better symbols of fidelity than ducks, and in Java it is geese, not ducks, that are given to newlyweds (and real geese, not wooden carvings).

Above: Carved Korean ducks are given as a wedding gift – they represent lasting devotion.

of a gander (although in some interpretations he rides a swan). The migrations of geese have led to ideas that they can foretell the weather. In ancient Nordic culture, people would examine the bones of a goose killed in autumn – dark spots on the sternum's keel indicated a severe winter was to come.

Ducks and geese have large families and are not shy about their courtship and mating behaviour, so a link with fertility is unsurprising – in the UK as well as elsewhere. The children's rhyme about Goosey Goosey Gander and his discovery of a badly behaved old man in 'my lady's chamber' seems to hint at frisky activity. 'Goose' is an old word for a prostitute and is sometimes still used to mean a salacious pinch on the bum. In a more maternal role, Mother Goose was the imaginary author of Charles Perrault's 17th-century fairy tales, a collection that included such classics as 'Cinderella', 'Sleeping Beauty' and 'Little Red Riding Hood'.

Ducks and geese in the media

Ducks are often the go-to comical animal of choice, and with good reason. Their squat frame, waddling gait and quacking voice all hit our funny bone, from our earliest years. Children's book, film and TV characters, like Disney's loud-mouthed Donald Duck and his extended family, and Warner Bros.' irascible but hapless Daffy Duck, are enduringly popular figures.

The typical fictional duck is brash, often grumpy and not very bright. One of the most famous in children's literature, Jemima Puddle-Duck, is a good-natured character but amazingly foolish, accepting the invitation of a hungry fox to come and nest in his shed. Beatrix Potter's 1908 tale concludes with a lucky rescue for Jemima. The duck in Sergei Prokofiev's 1936 musical tale *Peter and the Wolf* is less fortunate. She is caught and eaten by the wolf that comes to menace Peter and his grandfather, although all the other animals in the story escape thanks to young Peter's ingenuity. He manages to capture the wolf, ties it up and leads it to the zoo in a victory parade, during which the quacks of the unfortunate duck are just about audible from inside the wolf's belly – her voice is provided by a rather mournful oboe.

Right: In Prokofiev's much-adapted musical *Peter and the Wolf*, the wolf really wanted to eat Peter but instead made do with the duck.

Left: Designed by Walt Disney Productions, Donald Duck is a memorable and engaging cartoon character known for his unintelligible speech and exuberant personality.

Some more recent fictional ducks have had rather more edgy characters, such as Count Duckula, a duck who is also a vampire (albeit not a very good one) who began life as a semi-villain in the popular 1980s BBC animated series *Danger Mouse*, before starring in his own spin-off series. The character Duck in *Sarah and Duck*, an animated series for small children first broadcast on CBeebies in 2013, is a noisy, hyperactive male Mallard who accompanies the young heroine everywhere and is bright enough to open doors, take part in the children's games and play musical instruments.

Fictional geese are rarer than ducks. The best known is perhaps the Snow Goose from the 1940 novella of the same name by Paul Gallico. A lost migrant Snow Goose is shot and injured as it flies over the marshes of Essex. It is rescued by a young girl, Fritha, who takes it to local artist Philip Rhayader, a recluse who lives in an old lighthouse in the marshland. He nurses the bird back to health and sets it free, and it returns to visit him and Fritha on its migration over the coming years until Rhayader's death. The unashamedly sentimental tale has been narrated on air and adapted many times since its publication, remaining probably the best known of Paul Gallico's many works.

Peter Scott

No one individual has done more to protect and promote wildfowl in the UK than Sir Peter Scott. The only son of polar explorer Robert Falcon Scott, Peter was born in 1909 and from an early age had a passion for wildlife and also for art. He was just 26 when he had his first exhibition of bird paintings in London, many featuring flights of wild ducks or geese over wetland landscapes. An early interest in wildfowling was soon to change into a passion for wildfowl conservation.

In 1946, Scott founded the Severn Wildfowl Trust at Slimbridge in Gloucestershire, and set about captive breeding the endangered Hawaiian Goose (*Branta sandvicensis*) there. Some years later, captive-bred birds were returned to some of the Hawaiian Islands, bolstering the wild population and saving the species from extinction. The Severn Wildfowl Trust evolved into the Wildfowl & Wetlands Trust, acquiring nine more wetland sites in the UK. The WWT centres are nature reserves for wildlife, but also captive-breeding spaces for endangered birds from elsewhere in the world.

Above: Sir Peter Scott was a talented bird artist as well as a very committed conservationist.

Their work today includes saving the endangered Madagascar Pochard (*Aythya innotata*), both through captive breeding and by restoring suitable habitat in Madagascar for the captive-bred birds.

Above: The Visitor Centre at Slimbridge Wildfowl & Wetlands Trust was founded by Peter Scott, who believed that everyone should have the opportunity to get closer to nature.

Domestication

In the UK, two of our native species are fully domesticated – the Greylag Goose and the Mallard. They are joined in many British farmyards and smallholdings by two domesticated non-native species, the Swan Goose and the Muscovy Duck, which originate in eastern Asia and Central and South America, respectively.

We domesticate animals that are useful to us, and in the case of ducks and geese the uses are manifold – we can eat them and their eggs, stuff pillows with their soft down feathers and make pens from their strong flight feathers, and guard our property with them or just keep them as companion animals. Some countries still even practise the blood sport of 'goose-fighting', where aggressive ganders are pitted against each other for 'entertainment'.

Through selective breeding, we have developed distinct forms or breeds of domestic ducks and geese (especially Mallards) to suit our needs. For example, the Rouen, bred for meat, can weigh more than twice as much as a

Below: The four wildfowl species most common in captivity, and considered fully domesticated, are (clockwise from top left): the Mallard, Greylag Goose, Swan Goose and Muscovy Duck.

Above: Selective breeding of domestic birds can produce forms that are very different to the wild type. For example, the Indian Runner breed of domestic Mallard (left), and the Sebastopol breed of domestic Greylag Goose (right).

wild Mallard. The Indian Runner, with its distinctive and comical upright stance, is fast on its feet and easier to 'herd' than more waddling duck breeds. The charming little Call Duck was developed initially to be highly vocal and its calls were used to attract wild Mallards, but today it is kept mainly as a pet. Goose breeds include the large Emden, which can tip the scales at 14kg (30lb), and the bizarre ornamental Sebastopol goose, which has curly feathers on its body.

Right: Goose fighting is a popular sport in parts of Russia.

Colourful curiosities

Selective breeding of domestic animals can be used to preserve and combine unusual colour forms. Genetic mutations that cause unusual plumage coloration occur randomly and naturally, and the new colours can be propagated in the gene pool by back-crossing the odd-coloured birds' descendants. It is common to see domestic Mallards that are white, piebald or greenish black, or have diluted tones, while domestic geese are often white or piebald. The combination of strange colours and an unusual body shape make them look very different to their wild ancestors.

Because many domestic ducks and geese are abandoned at ponds, lakes and canals by disenchanted would-be poultry keepers, you will often find these oddities rubbing shoulders with normal-looking wildfowl, and they can cause much confusion to beginner birdwatchers. There is one infallible clue though, in male ducks at least – the presence of tightly curled-up central tail feathers will always indicate that your mystery bird, however peculiar looking, is a Mallard!

Above: A white domestic Greylag Goose rests happily among wild birds on the shores of a UK lake.

Watching Ducks and Geese

Wherever you are in the UK, there is probably a great site not far from your home where you can watch ducks and geese. Even in town centres, parklands with lakes can attract a surprising range of species, especially in winter. Seeing our less common and more localised species will probably involve some special excursions, but ducks and geese are great wanderers, and encountering a scarce or rare species can theoretically happen almost anywhere.

Where to watch wildfowl

Any kind of wetland habitat can be good for ducks, and geese are often encountered some distance from water, feeding on open fields. Winter is always the best season to look for an array of species, as so many of our wildfowl are winter visitors only (see the table on pages 117–19), but there is usually something to see at the best sites throughout the year.

Park lakes are a good starting point – most of us have one down the road. Here you will find Mallards and probably

Opposite: Greylag Geese, originally of captive origin but now fully naturalised, can be found in many areas of England, including the Lake District.

Left: Mixed flocks of wildfowl can be found on most large lowland lakes in the UK.

Above: To observe truly wild Greylag Geese in Britain, you will need to visit Scotland in winter.

some Tufted Ducks year-round. There may also be feral Greylag and Canada Geese, and if you are in the south you could also see Mandarins, Egyptian Geese and perhaps Red-crested Pochards. Winter could bring a few Pochards.

Most wild ducks and geese are too shy to feel comfortable close to people, but if you have a nature reserve nearby with some open water, you should see some of these shyer species as well. In most inland parts of the country, lowland and well-vegetated lakes will attract Teals, Gadwalls and Shovelers, and perhaps also Wigeons, Pintails and Goosanders. The closer your lake is to the coast, the higher the chances of seeing something more unusual – a Goldeneye or a seaduck perhaps.

Upland lakes and rivers in northern and western parts of the UK are the places to search for breeding Goosanders and Goldeneyes in summer. Lowland rivers will attract a few Mallards and perhaps other dabbling ducks, but it is their floodplains that are of more interest. Wet grazing marsh is a key winter habitat for Wigeons and also wild geese such as White-fronts, and in summer these areas attract breeding Garganeys. Some wild geese visit arable farmland, while estuaries are used by Brent Geese. Out at sea, look for Eiders, Long-tailed Ducks and Common and Velvet scoters in sheltered bays – the further north you are, the better the chances.

The table on the next three pages summarises the best habitats and regions in summer and winter for seeing our regularly occurring ducks and geese.

Species	UK habitats		UK distribution	
	breeding	migration and winter	breeding	migration and winter
GEESE				
Canada Goose *Branta canadensis*	Lowland urban and rural lakes and grassland	As for breeding habitats	Throughout England and Wales, parts of western and southern Scotland, and eastern Northern Ireland	As for breeding distribution
Brent Goose *Branta bernicla*	N/A	Muddy estuaries, saltmarshes, sheltered bays and coastal farmland	N/A	Widespread. Dark-bellied birds – The Wash, Kent and Hampshire; light-bellied – Strangford Lough and Lough Foyle in Northern Ireland, and Lindisfarne in Northumberland
Barnacle Goose *Branta leucopsis*	N/A	Coastal grassland and farmland	N/A	Islay and the Solway Firth
Greylag Goose *Anser anser*	Lowland urban and rural lakes, farmland and grassland	As for breeding habitats	Feral birds are widespread but scarcer in central and south-western regions	Wild migrants winter mainly in eastern Scotland
Pink-footed Goose *Anser brachyrhynchus*	N/A	Arable farmland and damp grassland, roosting on estuaries	N/A	Widespread, especially around The Wash, Solway Firth, Ribble estuary and eastern Scotland
Taiga Bean Goose *Anser fabalis*	N/A	Arable farmland and damp grassland	N/A	Yare Valley in Norfolk, and Falkirk in Scotland
Tundra Bean Goose *Anser serrirostris*	N/A	Arable farmland and damp grassland	N/A	Most frequent near the coast in eastern England
White-fronted Goose *Anser albifrons*	N/A	Arable farmland and damp grassland	N/A	Quite widespread, especially around the Severn and Swale estuaries, and western Scotland
DUCKS AND SHELDUCKS				
Egyptian Goose *Alopochen aegyptiaca*	Parkland and lowland lakes with fields and trees nearby	As for breeding habitats	Southern and eastern England, including the London area	As for breeding distribution
Shelduck *Tadorna tadorna*	Sheltered, undisturbed coastlines, including lagoons, and gravelly riversides	As for breeding habitats, plus muddy estuaries and sometimes inland lakes	Throughout coastal UK	As for breeding distribution, but more widespread, occurring further inland

Species	UK habitats		UK distribution	
	breeding	migration and winter	breeding	migration and winter
Mallard *Anas platyrhynchos*	Parks, gardens, lakes, rivers, marshland and anywhere else with fresh water	As for breeding habitats	Throughout the UK	As for breeding distribution
Pintail *Anas acuta*	Coastal wetlands with sparse vegetation	Larger lakes and lagoons, and wet marshland	Rare and sporadic, most often breeding in northern Scotland and eastern England	Quite widespread around coasts of England, Wales, Northern Ireland and eastern Scotland
Teal *Anas crecca*	Marshland with pools and ditches	Marshland and well-vegetated lakes and pools	Widely but sparsely distributed in the wider countryside, scarce in south-western and central England	Throughout lowland UK
Wigeon *Mareca penelope*	Undisturbed upland areas with small lakes	Lakes; also damp meadow-land, sometimes by the sea	Sparsely distributed in northern Scotland and northern England	Widespread throughout the UK, especially near coasts and estuaries
Gadwall *Mareca strepera*	Well-vegetated lakes	Parks, lakes, lagoons and marshland	Much of lowland central, southern and western England; more sparsely distributed near coasts in Wales, eastern Scotland and eastern Northern Ireland	As for breeding distribution but more widespread
Shoveler *Spatula clypeata*	Marshland with fresh water	Marshland, lakes and lagoons	Most common in eastern and northern England; sparsely distributed in Scotland and Northern Ireland	Widespread in lowland areas in England, and near the coast in Wales, Scotland and Northern Ireland
Garganey *Spatula querquedula*	Lush grassy marshland with pools and ditches	N/A	Mainly eastern and southern coasts of England	A little more widespread on autumn migration
Tufted Duck *Aythya fuligula*	Well-vegetated lakes	Lagoons, lakes in towns and the countryside, and reservoirs	Throughout England, and more sparsely distributed and coastal in Scotland, Wales and Northern Ireland	As for breeding distribution, but more widespread
Pochard *Aythya ferina*	Well-vegetated lakes	Lagoons, lakes in towns and the countryside, and reservoirs	Coasts and estuaries in eastern England, and sparsely distributed in Wales, Northern Ireland and southern Scotland	Quite widespread throughout the UK, especially near the coast
Scaup *Aythya marila*	N/A	Offshore in sheltered bays and estuaries, and on deeper lakes and lagoons near the coast	N/A	Widespread, most numerous around the Dee estuary and major Scottish estuaries

Species	UK habitats		UK distribution	
	breeding	migration and winter	breeding	migration and winter
Red-crested Pochard *Netta rufina*	Lakes, often in parkland	As for breeding habitats	Scarce, occurring mainly in south-eastern England	As for breeding distribution
Mandarin *Aix galericulata*	Lakes and slow-flowing rivers in or close to woodland	As for breeding habitats	Patchily distributed in lowland England (especially the south), and at a few scattered sites in Wales, Scotland and Northern Ireland	As for breeding distribution
Goosander *Mergus merganser*	Fast-flowing rivers and larger streams in uplands	As for breeding habitats, but also on lowland rivers and lakes	Upland northern and western England, upland Wales and northern Scotland	Much more widespread, found across most of the UK, although mostly near the coast in southern and eastern England
Red-breasted Merganser *Mergus serrator*	Upland rivers and sea coasts	As for breeding, but also offshore in bays and estuaries, and on coastal lagoons	Mainly in western and northern Scotland, north-west England and Wales, and coastal Northern Ireland	As for breeding distribution, but also occurs along eastern and southern coasts
Smew *Mergellus albellus*	N/A	Deep coastal lagoons and reservoirs	N/A	Potentially widespread but always scarce, most frequent in south-east England and on eastern coasts
Long-tailed Duck *Clangula hyemalis*	N/A	Offshore in sheltered waters, sometimes on coastal lakes, rarely inland	N/A	Most frequent off the east coast of England and Scotland and in the Irish Sea
Eider *Somateria mollissima*	Undisturbed rocky sea coasts and islands	Offshore in sheltered bays and estuaries	Around the coast of Scotland and northern England	Offshore around the whole of the UK
Common Scoter *Melanitta nigra*	Open upland moors with small pools	Offshore, mainly in sheltered bays and estuaries	Very sparsely distributed in northern Scotland	Offshore all around the UK, most common off Norfolk and in the Moray Firth, and in Carmarthen Bay and Cardigan Bay in Wales
Velvet Scoter *Melanitta fusca*	N/A	Offshore, mainly in sheltered bays and estuaries	N/A	Rare; may be found around much of the coastline, but particularly off north-eastern England and south-eastern Scotland
Goldeneye *Bucephala clangula*	Slow-flowing upland rivers and forest lakes	Larger and deeper lakes and lagoons, sometimes offshore	Sparsely in forested parts of northern Scotland	Widespread, especially close to the coast

How to watch wildfowl

As soon as your wildfowl watching takes you beyond the local park, you will need binoculars. Wild ducks and geese tend to be wary and are often on water, so cannot easily be approached even if they are not too fearful. You will also find a telescope very useful at some sites, and for seawatching. Even when you are out in the countryside far from any water, you may see migrating wildfowl flying overhead, and these flocks are likely to go unidentified if you are not carrying binoculars.

When you are learning to identify ducks, it is easiest to get to grips with breeding-plumaged males first of all. Most will be in full breeding plumage between November and May. Females will usually associate most closely with males of their own species, so you can compare them side by side. Focusing on details of shape and pattern will help you get more skilled at picking out birds of different species in poor light over a long distance, and also at identifying males in eclipse plumage and young birds in their female-like juvenile plumage. Geese lack bewildering plumage variation and most species are quite distinctive, but telling apart the 'grey geese' is not always straightforward. Make a point of checking bill colour and pattern, as this is where the four potential confusion species are reliably distinct from one another.

Below: Nature reserves, such as RSPB Greylake in Somerset, can attract a great range of wildfowl species including Mallards and Pochards.

Found a rarity?

If you see a duck or goose that you cannot identify, first of all eliminate the possibility that it is an escaped domestic bird. Many peculiar-looking ducks seen in the wild are actually domestic Mallards. If your bird strikes you as a strange version of a familiar species, it may well be a hybrid, most of which show clear traits of at least one of their parents.

If you do identify your mystery bird as belonging to a species that does not usually occur in the UK, there are two possibilities. Either it is an escapee from a zoo or exotic bird collection, or it is a wild bird that has strayed off course. Some unusual species are certain to be escapees – this category would include any bird native to South America or Australia, for example. For individuals of species native to North America or Europe, the picture is less clear – your bird could be an escapee or a wild wanderer. Take photos of the bird if you can, and try to see if it is wearing a leg ring. Share your sighting with other birdwatchers, or report it to your county bird recorder. If you are on a nature reserve, you could report it to the site warden.

Above: Unusual wildfowl found in the UK, including Bar-headed Geese, are usually escapees from captivity.

Observing duck and goose behaviour is always interesting, particularly their courtship displays and parenting behaviour. Any wetland nature reserve with hides overlooking water offers the chance of watching wildfowl without the risk of disturbing them. While you are watching, you may also see a wide range of other wildlife, including many birds but also mammals, possibly fish and amphibians, and, in the warmer months, a great variety of insects. Our wetlands are among the richest and most important of all our wild habitats, and their ecology, both above and below the surface, is wonderfully diverse. A love of ducks, geese and other wetland birds is the starting point to learning about how their watery world fosters and sustains life of all kinds.

Above: Observing flocks of wild geese on the move is one of the great joys of wildlife-watching.

Glossary

Aberrant A bird with non-typical plumage colours or some other unusual physical trait.

Banding *See* 'ringing'.

Bill Another word for a bird's beak, used for all species of birds but particularly for ducks, geese and swans.

Brood patch An area of bare skin on the belly that develops during breeding, to help make incubation more efficient.

Conservation The set of measures and actions applied to attempt to halt and reverse the decline of a species and to safeguard its habitats.

Courtship Ritualised behaviours performed by the male and female together, to help form or consolidate a pair bond.

Crèche A large assembly of baby birds from multiple families, attended by a few adults.

Dabbling duck A duck that takes most of its food at the surface of the water.

Display Postures and movements made mainly by male ducks, to show off particular plumage features to a prospective mate or to rivals.

Diving duck A duck that dives underwater to find food.

Domesticated A species that has been kept and bred in captivity for many generations.

Down Soft, fluffy feathers, some of which are shed to line the nest (and create a brood patch).

Eclipse A temporary female-like body plumage developed in many male ducks during the summer moult, for camouflage to help protect them while they are replacing their flight feathers.

Escapee An individual bird that has escaped from captivity (and can usually be recognised as such – for example, by a leg ring).

Estuary The mouth of a river, with tidal salty water and a muddy shoreline.

Feral A bird that has escaped from captivity, or a descendant of these birds.

Fledging When a young bird begins to fly. In ducks, geese and other precocial birds, this happens weeks after leaving the nest.

Flight feathers The long wing feathers – the primaries (on the 'hand', or outermost part, of the wing) and secondaries (on the 'arm').

Habitat The place where a species lives, including its resident ecosystem of plants and other animals.

Hybrid A bird whose parents are two different species; hybridisation is more frequent in ducks and geese than in other groups of birds.

Incubation Applying constant warmth to the eggs through body heat, to enable the embryos within to develop.

Lamellae Projections on the cutting edge of the bill. They may be hair-like, sharp or blunt, depending on the bird's mode of feeding.

Migration A regular seasonal journey a bird makes between its breeding areas and wintering areas. See also 'moult migration'.

Migratory Any bird that habitually migrates, in at least part of its range.

Moult migration A journey made by some ducks after breeding, often to traditional sites where they will undergo their annual moult.

Nest box An artificial structure designed for birds to nest inside. Often used by species like Goldeneye that nest in tree-holes.

Non-native Any animal living in an area where it does not naturally occur, following introduction there (deliberate or accidental) by people.

Precocial A young bird that hatches in a well-developed state, with a full covering of down, open eyes, and the ability to move freely and feed itself.

Predator An animal that hunts and kills other animals.

Preen gland A gland at the base of the tail that secretes preen oil. The bird applies this to the feathers to discourage parasites, help with waterproofing and produce certain scents. Also called the uropygial gland.

Primaries The longest flight feathers, forming the trailing edge of the 'hand' (outermost part) of the wing.

Quarry A species of animal that can legally be killed for food in a particular country.

Resident A bird that does not migrate.

Ringing Placing uniquely numbered rings on a bird's leg, for future identification. Also called banding.

Sawbill A duck of the genus *Mergus* or *Mergellus*, which specialises in pursuit-hunting fish underwater, and has a serrated bill to grip its prey.

Seaduck A duck that lives on the open sea when not breeding.

Seawatching Watching birds that are passing by out at sea. Seawatching usually requires the use of a telescope.

Secondaries The inner flight feathers, forming the trailing edge of the 'arm' of the wing.

Shelduck Birds of the subfamily Tadorninae, intermediate between ducks and geese.

Survey A study that involves counting the number of wild birds present at a particular site at a certain time.

Uropygial gland *See* 'preen gland'.

Vagrant An individual bird that has strayed a long way beyond its normal distribution.

Waterfowl Any bird that lives on or around fresh water.

Webbing The membrane of skin that stretches between the toes in many swimming birds.

Wetland Any habitat with some fresh water – includes lakes and rivers, marshland, wet meadows, reedbeds and estuaries.

Wildfowl Birds of the family Anatidae – the ducks, geese and swans.

Acknowledgements

Thank you to Julie Bailey for asking me to write this book about one of my favourite groups of wild birds, and to Molly Arnold and Jenny Campbell for managing the project through the design stages. I am grateful to Susi Bailey for copy-editing my manuscript, Susan McIntyre for creating the page layout, and Lucy Beevor for proofreading. Many photographers have supplied the array of images that bring these pages to life, and I would like to thank them all for their enthusiasm and talent; likewise, I thank the many researchers and conservationists whose findings I have drawn upon to tell the stories of our wildfowl, both here and abroad. Finally, thanks go to my friends and family for their constant support, and especially to Alex for his helpful comments on the text and for all the cups of tea.

Further Reading and Resources

Books

Cabot, D. 2009. *Wildfowl*. New Naturalist Library. HarperCollins, London.

Cocker, M. & Mabey, R. 2005. *Birds Britannica*. Chatto and Windus, London.

Humble, K. & McGill, M. 2011. *Watching Waterbirds with Kate Humble and Martin McGill*. A&C Black, London.

Huxley, E. 1994. *Peter Scott: Painter and Naturalist*. Faber & Faber, London.

Kear, J. (ed.). 2005. *Ducks, Geese, and Swans: Anseriformes. Vols 1 and 2*. Bird Families of the World. Oxford University Press, Oxford.

Ogilvie, M. A. & Young, S. 1998. *Photographic Handbook of the Wildfowl of the World*. New Holland Publishers Ltd, London.

Reeber, S. 2015. *Wildfowl of Europe, Asia and North America*. Helm Identification Guides. Christopher Helm, London.

Websites

British Trust for Ornithology (BTO)
bto.org
Concerned with accurately monitoring the UK's birdlife, including running a range of surveys to which ordinary birdwatchers can contribute.

The Department for Environment, Food and Rural Affairs (Defra)
Tel: 03459 335577
Any dead wildfowl found in the UK should be reported to Defra on the number above.

IUCN Red List
iucnredlist.org
Collates and maintains conservation information for all life on Earth on both local and global levels.

Royal Society for the Protection of Birds (RSPB)
rspb.org.uk
The UK's leading wildlife conservation charity, with a website full of information on all aspects of nature and conservation.

Wildfowl & Wetlands Trust (WWT)
wwt.org.uk
A conservation body founded by Sir Peter Scott, which now manages land and captive-breeding projects to protect threatened wildfowl and other aquatic life.

Xeno-canto
xeno-canto.org
A huge database of bird songs and calls.

The following websites all sell grains and pellet food suitable for wild ducks and geese:

Ark Wildlife
arkwildlife.co.uk

Busy Beaks
busybeaks.co.uk

CJ Wildlife
birdfood.co.uk/duck-and-swan-food

Image Credits

Bloomsbury Publishing would like to thank the following for providing photographs and permission to produce copyright material. While every effort has been made to trace and acknowledge all copyright holders, we would like to apologise for any errors or omissions and invite readers to inform us so that corrections can be made in any future editions of the book.

Key t = top; l = left; r = right; tl = top left; tr = top right; c = centre; cl = centre left; cr = centre right; b = bottom; bl = bottom left; br = bottom right

AL = Alamy; G = Getty Images; IS = iStock; NPL = Nature Picture Library; RS = RSPB Images; SH = Shutterstock

Front cover t Steve Round/RS, b Kevin Sawfor/RS; **spine** Savo Ilic/SS; **back cover** t Henri_Lehtola/SS, b Giedriius/SS; **1** Wildlife World/SS; **3** David Tipling/birdphoto.co.uk; **4** wonderful-Earth.net/AL; **5** Wang LiQiang/SS; **6** t Steve Oehlenschlager/SS, b Johnwoodkim/SS; **7** l crbellette/IS, r Marianne Taylor; **8** tl Robert L Kothenbeutel/SS, tr Martin Fowler/SS, b Antero Topp/SS; **9** Francis Bossé/SS; **11** Simonas Minkevicius/SS; **12** Erni/SS; **13** tl Erni/SS, tr Helen J Davies/SS, b atosf/IS; **16** Mariia Zaporozhtseva/SS; **17** l godi photo/SS, r WildlifeWorld/SS; **18** t Mirko Rosenau/SS, b Ponderful Pictures/SS; **19** t Mark Richardson Imaging, b Marcin Perkowski/SS; **20** t Wolfgang Kruck/SS, c Andrew Harker/SS, b Erni/SS; **21** t Natural Imaging/SS, c Erni/SS, b Dennis Jacobsen/SS; **22** Marianne Taylor; **24** Smokeybjb/Wikimedia Commons; **25** l Apokryltaros/Wikimedia Commons, r Apokryltaros/Wikimedia Commons; **26** Erni/SS; **27** t Betty4240/IS, b kitschinc/SS; **28** Chris_M_Rabe/SS; **29** tl Christophe Mischke/SS, tr Tom Reichner/SS, bl Agami Photo Agency/SS, br Agami Photo Agency/SS; **30** t Nicram Sabod/SS, b Minden Pictures/AL; **31** l Rudmer Zwerver/SS, r David Tipling/birdphoto.co.uk; **32** Marianne Taylor; **33** t Marianne Taylor, b Mark Richardson Imaging; **34** t Marianne Taylor, b Marianne Taylor; **35** David Tipling/birdphoto.co.uk; **36** Mark Richardson Imaging; **37** Marianne Taylor; **38** Smiller99/SS; **39** t Marianne Taylor, b VictorTyakht/IS; **40** Alfred Trunk/BIA/NPL; **41** l Marianne Taylor, r Marianne Taylor, b Simon Vasut/G; **42** Marianne Taylor; **43** David Tipling/birdphoto.co.uk; **44** David Hosking/AL; **45** imageBROKER/AL; **46** blickwinkel/AL; **47** Agnieszka Bacal/SS; **48** t Marianne Taylor, b Vishnevskiy Vasily/SS; **49** Sheila Fitzgerald/SS; **50** Anneka/SS; **51** Marianne Taylor; **52** Minden Pictures/AL; **53** Jeff Vanuga/NPL; **54** wonderful-Earth.net/AL; **55** t Tom Reichner/SS, b Andrew M. Allport/SS; **56** Chris O'Reilly/RS; **57** Marianne Taylor; **58** Marianne Taylor; **59** t David Chapman/AL, b Marianne Taylor; **60** Marianne Taylor; **61** t Erni/SS, b MichaelGrantBirds/AL; **62** Anand B/500px/G; **63** John Keeble/Contributor/G; **64** t Marianne Taylor, b Adrian Sherratt/AL; **65** David Tipling/birdphoto.co.uk; **66** Ann and Steve Toon/AL; **67** Gerrit Vyn/NPL; **68** Nick Upton/NPL; **69** Guy Edwardes/2020VISION/NPL; **70** tony mills/SS; **71** Graham Catley/AL; **72** t DAVID TIPLING/NPL, b Agami Photo Agency/SS; **73** Marianne Taylor; **74** t Donald M. Jones/NPL, b David Tipling/birdphoto.co.uk; **75** Marianne Taylor; **76** Markus Varesvuo/NPL; **77** David Tipling/birdphoto.co.uk; **78** David Tipling/birdphoto.co.uk; **79** David Tipling/birdphoto.co.uk; **80** Incredible Arctic/SS; **81** Kennerth Kullman/SS; **82** fotolincs/AL; **83** t Raffi Maghdessian/Aurora Photos/G, b Chaithanya Krishnan/G; **84** BIOSPHOTO/AL; **85** SanderMeertinsPhotography/SS; **86** David Tipling/birdphoto.co.uk; **87** Michael Schroeder/AL; **88** Jeff Vanuga/NPL; **89** Mark Richardson Imaging; **90** t Markus Varesvuo/NPL, b Westend61/G; **91** Steven J. Kazlowski/AL; **92** Steve Oehlenschlager/SS; **93** t Robin Chittenden/NPL, b Marianne Taylor; **94** Ingrid Maasik/SS; **95** Markus Halbmair/SS; **96** Julia McGee/SS; **97** t bchyla/SS, b YuryKara/SS; **98** AGAMI Photo Agency/AL; **99** Giedriius/SS; **100** Erni/SS; **101** Andrew Mason/RS; **102** David Tipling/birdphoto.co.uk; **104** DaisyPhotography/AL; **105** PRISMA ARCHIVO/AL; **106** t Hemis/AL, b Historic Collection/AL; **107** cabecademarmore/AL; **108** Photo 12/AL; **109** B Christopher/AL; **110** t Pictorial Press Ltd/AL, b foto-zone/AL; **111** tl Marianne Taylor, tr Martin Fowler/SS, bl Tom Meaker/SS, br SunflowerMomma/SS; **112** tl wrangel/IS, tr KiraVolkov/IS, b Nikolay Vinokurov/AL; **113** Marianne Taylor; **114** Mark Richardson Imaging; **115** Nick Upton/NPL; **116** Marianne Taylor; **120** David Kjaer/RS; **121** Marianne Taylor; **122** David Kjaer/NPL.

Index